Joseph William Richards

Aluminum

Its History, Occurrence, Properties, Metallurgy and Application....

Joseph William Richards

Aluminum

Its History, Occurrence, Properties, Metallurgy and Application....

ISBN/EAN: 9783337203719

Printed in Europe, USA, Canada, Australia, Japan

Cover: Foto ©berggeist007 / pixelio.de

More available books at **www.hansebooks.com**

ALUMINIUM:

ITS HISTORY, OCCURRENCE, PROPERTIES,
METALLURGY AND APPLICATIONS,
INCLUDING ITS ALLOYS.

BY

JOSEPH W. RICHARDS, A.C.,

CHEMIST AND PRACTICAL METALLURGIST; MEMBER OF THE DEUTSCHE
CHEMISCHE GESELLSCHAFT.

ILLUSTRATED BY SIXTEEN ENGRAVINGS.

PHILADELPHIA:
HENRY CAREY BAIRD & CO.,
INDUSTRIAL PUBLISHERS, BOOKSELLERS AND IMPORTERS,
810 WALNUT STREET.

LONDON:
SAMPSON LOW, MARSTON, SEARLE & RIVINGTON,
CROWN BUILDINGS, 188 FLEET STREET.
1887.

COPYRIGHT BY
JOSEPH W. RICHARDS,
1886.

COLLINS PRINTING HOUSE,
705 Jayne Street.

PREFACE.

No apology is necessary in presenting a work on aluminium in English. In 1858 Tissier Bros. published in France a small book on the subject. H. St. Claire Deville, the originator of the aluminium industry, published a treatise, also in French, in 1859. Deville's book is still the standard on the subject. Until December, 1885, we have an intermission, and then a work by Dr. Mierzinski, forming one of Hartleben's Chemisch-Technische Bibliothek, which is a fair presentation of the industry up to about 1883, this being a German contribution. Probably because the English-speaking people have taken comparatively little hand in this subject we find no systematic treatise on aluminium in our language. The present work aims to present the subject in its entirety to the English reader.

Tissier, Deville, Mierzinski, and the German, French, and English scientific periodicals have been freely consulted and extracted from, full credit being given in each case to the author or journal. As this art has of late advanced so rapidly it has been a special aim to give everything that has been printed up to the time of publication.

The different parts of the work are arranged in what seemed their logical order, corresponding closely to that followed by Deville. The Appendix contains an account of laboratory experiments, etc., several of which, it is trusted, may be of value.

In conclusion, the author wishes to thank the faculty of his "Alma Mater," Lehigh University, for their permission to use his Thesis on Aluminium as the basis of this treatise; also, to acknowledge his indebtedness to Dr. Wm. H. Greene, of Philadelphia, for assistance rendered in the preparation of the work for the press.

J. W. R.

PHILADELPHIA, November 25, 1886.

LIST OF REFERENCES.

Tissier	Recherche de l'Aluminium. C. & H. Tissier. Paris, 1858.
Deville	De l'Aluminium. H. St. Claire Deville. Paris, 1859.
Watts	Watts's Dictionary of Chemistry, vol. i.
Mierzinski	Die Fabrikation des Aluminiums. Dr. Mierzinski. Vienna, 1885.
Compt. Rend.	Comptes Rendus de les Sciences de l'Académie. Paris.
Wagner's Jahresb.	Wagner's Jahresbericht der Chemische Technologie.
Phil. Mag.	The London and Edinburgh Philosophical Magazine.
Mon. Scientif.	Le Moniteur Scientifique. Dr. Quesnesville.
Fremy	Encyclopédie Chemique. Fremy. Paris, 1883.
Dingl. Joul.	Dingler's Polytechnisches Journal.
Pogg. Ann.	Poggendorff's Annalen.
Jrnl. der Pharm.	Journal der Pharmacie.
Bull. de la Soc. Chem.	Bulletin de la Société Chemique de Paris.
Sci. Am. (Suppl.)	Scientific American (Supplement).
Eng. and Mng. Jrnl.	The Engineering and Mining Journal.
Chem. News	The Chemical News.
Jahresb. der Chem.	Jahresbericht ueber die Fortschritte der Chemie.

FORMULÆ.

Al . . . . . Aluminium.
Al²O³ . . . Alumina.
Al²Cl⁶ . . . Aluminium chloride.
K . . . . . Potassium.
KOH . . . Caustic Potash.
KCl . . . Potassium chloride.
Na . . . . Sodium.
Al²Cl⁶.2NaCl. Aluminium-sodium double chloride.
Si . . . . Silicon.
Fe . . . . Iron.
Cu . . . . Copper.

TEMPERATURES.

Unless stated otherwise, all temperatures given are in Centigrade degrees.

CONTENTS.

PART I.

HISTORY OF ALUMINIUM.

	PAGE
Lavoisier's suggestion of the existence of metallic bases of the earths and alkalies; Researches in preparation of aluminium, by Davy, Oerstedt, and Wöhler	25
Isolation of aluminium, by H. St. Claire Deville, in 1854	26
Patronage of Emperor Napoleon III.; Aluminium at the Paris Exhibition, 1855; Its manufacture on a large scale at Glacière, Nanterre, and Salindres; Tissier Bros.' book on aluminium in 1858	28
Deville's book, 1859; History of the works near Rouen	29
Deville's explanation of the uses of the new metal	30
Alfred Monnier's production of aluminium at Camden, N. J., 1856	31
W. J. Taylor claiming the possible cost of aluminium at $1 per pound; Kerl and Stohman's résumé of the manufacture of aluminium up to 1874	32
Dr. Clemens Winckler's retrospect of the development of aluminium, 1879	33
Manufacture of aluminium in England, France, and Germany; Aluminium beams for balances, made by Sartorius of Göttingen; Difficulties in using aluminium for mathematical instruments; Action of molten aluminium upon earthen crucibles	35

CONTENTS.

	PAGE
Prices of aluminium and of aluminium bronze in France; Webster's aluminium works in England	36
Col. William Frishmuth's invention for producing aluminium at reduced cost; Opinion of his invention by Major Ricarde-Seaver, F.R.S.E.	37
Col. Frishmuth's works in Philadelphia; Aluminium casting for the Washington Monument, made by Col. Frishmuth; Census report of his annual production; His price in bars	39
Imports and consumption of aluminium in the United States from 1870 to 1884; Its production in Philadelphia by Col. Frishmuth in 1883, 1884	40
Cowle's process for making aluminium bronze at Cleveland, Ohio; Present state of aluminium industry as described by Prof. Charles F. Mabery, of Cleveland, Ohio, and Dr. T. Sterry Hunt, of Montreal	41

PART II.

OCCURRENCE OF ALUMINIUM IN NATURE.

Combinations of aluminium with oxygen, alkalies, and acids, etc.; Formulæ of aluminium silicates	43
Appearance of most of the aluminium compounds; Formulæ of some of the precious stones	44
Minerals most used for producing aluminium; Bauxite	45
Analyses of bauxite	46
Cryolite; Where found, description, and general uses; Its importation by the Pennsylvania Salt Co., of Philadelphia; Native cryolite in the United States	48
Imports of cryolite into the United States; Corundum; Its great source of supply	49
Probable sources of supplies of materials for production of aluminium in the United States and Great Britain	50

CONTENTS. ix

PART III.

PHYSICAL PROPERTIES OF ALUMINIUM.

PAGE

Table of analyses of commercial aluminium . . . 51
Free and combined silicon in aluminium; Gases in aluminium 52
Composition of the aluminium apex of the Washington Monument at Washington, D. C., cast by Col. Frishmuth; Color of aluminium; As described by Deville, Fremy, Mallet, and Mierzinski 53
Mat; As described by Deville, Mierzinski, and Bell Bros. 54
Polish and lustre; Processes for producing as given by Deville, Bell Bros. and Kerl and Stohman . . 55
Odor; As given by Deville and Watts 56
Taste—Deville; Malleability—Deville and Mallet on this subject 57
M. Degousse's success in beating aluminium into leaves; Substitution of aluminium for silver leaf; Kerl & Stohman on rolling and annealing aluminium . . 58
Bell Bros. on beating aluminium; Mierzinski on extensibility of aluminium 59
Aluminium leaf first made by C. Falk & Co., Vienna; Ductility; Drawing aluminium wire; Results obtained by Deville, Vangeois, and Bell Bros. . . . 60
Elasticity, Tenacity, Hardness—Deville, Wertheim, Mallet, Fremy; Kerl & Stohman on engraving aluminium 61
Mierzinski and W. H. Barlow on tensile strength; Tables; Comparative mechanical value of aluminium, steel, etc. 62
Table of strength of aluminium wire; Sonorousness; Results obtained by Deville and M. Lissajous in making bells and tuning-forks 63
Results obtained by Faraday and Watts; Density; Deville's table of comparison with other metals . . 64

Comparative value of equal volumes of aluminium and silver; Specific gravity of absolutely pure aluminium, Mallet; Fusibility; As given by Deville, Mallet, and Mierzinski 65
Fixity; As given by Deville, Watts, and Fremy; Electric conductivity; Results obtained by Deville and M. Buff 66
Comparison with copper and magnesium, Fremy; Thermal conductivity; Deville, Faraday, and Watts, etc., on this subject 67
Mierzinski, Calvert, and Johnson on this subject; Specific heat; Deville, Regnault, Paul Morin, Mallet, and Fremy on this subject 68
Magnetism; Deville, MM. Poggendorff and Reiss; Crystalline Form; Deville on this subject . . . 69

PART IV.

Chemical Properties of Aluminium.

Remark; Action of air; Deville's observations . . 70
Cupellation of aluminium; Observations of Wöhler, Peligot, Watts, etc. 71
Action of water; Deville on this subject . . . 72
Mierzinski and the Chemical News; Action of hydrogen sulphide and sulphur; Deville and Fremy . . 73
Sulphuric acid; Deville, M. de la Rive, and Fremy . 74
Nitric acid; Deville and M. Hulot; Hydrochloric acid; Deville, M. Favre, and others 75
Potash, soda, and lime; Deville, Mallet, Mierzinski . 77
Aqua ammonia; Deville and Wöhler; Organic acids, vinegar, etc.; Deville, M. Paul Morin; Use of Aluminium for culinary articles 78
Solutions of metallic salts; Precipitation of other metals by aluminium; Deville, Tissier, Paul Morin, Mourey, Christofle, Hulot 79

	PAGE
Mierzinski, Fremy and Watts	81
Nitre; Purification by nitre; Deville, Fremy, and Mierzinski on this subject	83
Silicates and borates; Action on glass and crucible clay; Deville and Tissier; Fluorspar; Tissier on its use as a flux	84
Phosphate of lime, Tissier on this subject; Sodium chloride and chlorides, Deville; Tissier on their use as fluxes	85
Metallic oxides; Tissier's experiments	86
Mierzinski; Békétoff's experiments; Animal matters; Deville and M. Charrière on the use of aluminium in surgery	87
Miscellaneous agents; Tissier and Mierzinski on this subject	88
General observations on the properties of aluminium, Deville	88

PART V.

METALLURGY OF ALUMINIUM.

Oerstedt's original paper on isolation of aluminium, 1824	90
Wöhler, the true discoverer of the metal; Wöhler's first paper	91
Wöhler's second paper	93
Deville's remarks on the metal obtained by Wöhler	95
Deville's improvement, 1854–55	96
Deville's apparatus at Javel and Glacière described and illustrated	98
Deville's experiments with sodium vapor	100
Reduction from cryolite; H. Rose's entire paper	103
Dr. Percy's investigations as laid before the Royal Institution	115
Allan Dick's paper, November, 1855	116

	PAGE
Deville's account of his researches	118
Wöhler's improvement on Deville's process; Watts on the reduction of cryolite; Gerhard's furnace	126
Watts's summary of the use of cryolite	127
General remarks	128

PART VI.

The Manufacture of Sodium.

Preliminary observations	130
Summary, taken principally from Mierzinski; Efforts of Davy, Gay Lussac, Thenard, Curaudau, Brünner, Donny, and Mareska	131
Donny and Mareska's condenser, illustrated; Deville's account of its operation	132
Object of and disadvantage in use of chalk; Preliminary calcination of the mixture; Illustration of the furnace.	133
Decomposition retorts, illustrated	134
Operation in the retorts	135
Deville, Rivot, and Tissier on the temperature	137
Wagner's improvement; Attempts to reduce potassium and sodium together	138
Weldon's calculation of the cost of sodium; Making of sodium in New York City, N. Y.; Castner's American patent process	139
Claims made in Castner's patent	141
Reduction of sodium by electricity, Mierzinski, Davy; Jablochoff's apparatus described and illustrated	142

PART VII.

Manufacture of Alumina.

Present state of the industry; Tilghman's process	144
Manufacture from cryolite; Dry way	146

CONTENTS. xiii

	PAGE
Thomson's furnace described and illustrated	147
Preference for mechanical furnaces as used in manufacture of soda, potash, etc.	148
Precipitation of solution of sodium aluminate, according to Löwig	151
Wet way	152
Manufacture from alum-stone or shales	153

PART VIII.

MANUFACTURE OF THE DOUBLE CHLORIDE OF ALUMINIUM AND SODIUM.

Preliminary remarks	154
Mierzinski, Deville, and M. Dullo on this subject	155
Manufacture by using hydrochloric acid and carbon disulphide	157

PART IX.

MANUFACTURE OF ALUMINIUM AT SALINDRES (GARD).

Aluminium as made by À. R. Pechiney & Co., successors to Henry Merle & Co.	158
Reactions involved in and outline of the process	158
Preparation of the aluminate of soda; Material used; Composition of mixture; Calcination, washing, filtering, with illustration of filtering apparatus	159
Preparation of the alumina; Description of precipitating tank and method of precipitation, washing, and drying, illustrated	163
Preparation of aluminium—sodium double chloride	166
Illustration of furnace; Mixing and shaping the charge; Condenser	167

2

	PAGE
Reduction of the double chloride by sodium; Illustration of furnace; Difficulties met; Method of charging, reducing, and running out	168
Average cost of manufacture at Salindres in 1872	172
Later improvements in Deville's process; Webster's process; History and description of the plant; Where its advantages lie; Utilization of bye products	173
Frishmuth's process; Patent claims	178
Other processes; Niewerth's method of reduction by nascent sodium; Grousillier's reduction under pressure	179

PART X.

REDUCTION OF ALUMINIUM BY OTHER REDUCING AGENTS THAN SODIUM.

Reduction by Cyanogen; Knowles's patent; Corbelli's patent	180
Deville's and Watts's comments; Reduction by hydrogen; Process of F. W. Gerhard; Comment by Watts	181
Reduction by carburetted hydrogen; Process of A. L. Fleury, of Boston	182
Petitjean's process	183
Reduction by double reaction; Processes of M. Comenge and Johnson	184
Process of Niewerth	185
Reduction by carbon and carbon dioxide; Process of J. Morris, of Uddington	187
Reduction by carbon; Article by M. Chapelle	188
Statement of G. W. Reinar; Cowles Bros.' process	189
Patent claim of Messrs. Cowles	190
Prof. Charles F. Mabery's official account of Cowles Bros.' process	191
Dr. T. Sterry Hunt's paper read before the American Institute of Mining Engineers	194

CONTENTS.

	PAGE
Dr. Hunt's address before the National Academy of Science	196
W. P. Thompson's complete description of the process	197
Illustrative description of the furnace; Mode of operating furnace, and improvements thereon; Amount reduced; Ores used	199
Reduction by iron; Lauterborn's process not new; Patents of F. Lauterborn and of H. Niewerth	206
Preparation of aluminium and sodium in the Bessemer converter; W. P. Thompson's experiments	207
Calvert and Johnson's experiments	209
Reports of Fremy, Watts, Benzon, Evrard	211
Silicon bronze, by Evrard; Ostberg's statement of the iron-aluminium alloy used in the mitis process; Reduction with copper; Calvert and Johnson's process	212
Kerl and Stohman's account of Benzon's process	213
Laboratory tests of this process; Reduction by zinc; Dullo's observations	214
Patent of M. N. Basset	215
Wedding's remarks on Basset's process	217
Kagensbusch's singular proposition; Fred'k J. Seymour's patent	218
Extraordinary claim in Seymour's second patent	220
American Aluminium Co., Detroit; Reduction by lead; Wilde's invention	221
Reduction by manganese; Claims of W. Weldon, Burstow, England; Reduction by electricity	222
Deville's account of the process	223
Sectional illustration of the crucibles	224
Bunsen and Deville on the subject	225
Mierzinski's practical remarks	226
Patented improvement by Richard Grätzel, Germany, illustrated	228
Duvivier's experiment with electric current	229

xvi CONTENTS.

PAGE

Kagensbusch's proposition; Gaudin's "economic" reduction of aluminium 230
Metals coated with aluminium by Thomas and Tilly; Depositing of aluminium by Corbelli and J. B. Thompson 231
Patented process by J. A. Jeancon; Experiments by M. A. Bertrand, C. Winkler, and Sprague . . . 232
Electrolyses of M. L. Senet, Gerhard, and Smith; Decomposition of a solution of alum by J. Braun; Moses G. Farmer's patent for obtaining aluminium . . 233
Mierzinski's denial of the successful deposition of aluminium from an aqueous solution of its salt; Aluminium and nickel plating at Frishmuth's works . . 234

PART XI.

WORKING IN ALUMINIUM.

Melting aluminium; Deville's instructions . . . 235
Kerl and Stohman's instructions 236
Mierzinski's instructions; Casting aluminium; Deville's instructions 237
Purification of aluminium; Freeing from slag, Deville . 238
Process of Paul Morin 239
Watts's suggestion; Freeing from impurities, Deville . 240
Mierzinski's recommendation 242
Buchner's treatment of commercial aluminium to eliminate silicon; Mallet's process of obtaining pure from commercial aluminium; Uses of aluminium . . 243
Aluminium plating and aluminium leaf 246
Aluminium coins; Soldering aluminium; Deville's views on 247
Hulot's process; Mourey's solder 248
Mierzinski's statements as to Mourey's solder; Improvements of Schwarz; Formulæ for these solders . . 249

CONTENTS. xvii

PAGE

Frishmuth's solders; Kerl and Stohman on Mourey's
solders, with formulæ 250
Process of Bell Bros. 252
Veneering with aluminium; Deville's account of - the
success of M. Sevrard in 1854 253
Dr. Clemens Winckler on this subject 254
Gilding and silvering aluminium; Failures of Deville
and Morin; Success of Mourey and Christofle . . 256
Watts, and Kerl and Stohman, on this subject . . 257.

PART XII.

ALLOYS OF ALUMINIUM.

General remarks, Mierzinski 258
Aluminium and silicon; Tissier and Deville . . . 259
Aluminium and mercury; Statements of Deville and
Watts; Aluminium amalgam made by Caillet with the
battery 261
Joules's method of electrolyzing 262
Properties of aluminium amalgam; Fremy, Tissier,
Gmelin on this subject 263
Aluminium and copper; Tissier Bros., 1858 . . . 264
Deville, 1859; Use of the alloy by Christofle; Alloy described by Debray; Composition of aluminium bronze 265
Properties of aluminium bronze; M. Lechatelier's table
of its strength; Experiments by A. Gordon . . 266
M. Boudaret on its malleability; Mierzinski on points to
be attended to in making the aluminium bronze . . 267
Directions to be observed in casting; Comparative
strength of the bronze 268
Hulot's solder; Fremy's instructions; Kerl and Stohman's
directions 269
Bronze for philosophical instruments; Specific gravity
and strength; Comparative strength of the alloys;
Their specific gravities 270

CONTENTS.

	PAGE
Melting point of 10 per cent. bronze; B. S. Procter's experiments	271
Thurston on the properties of aluminium bronze	272
Strange and Knight on the properties of aluminium bronze	273
Alloys made by Cowles Bros.	274
Strength of these alloys by the testing machine	276
Alloys of aluminium and copper with other metals; Neogen made by F. H. Sauvage	277
Minargent; P. Baudrin's alloy; James Webster's patent bronze	278
Phosphor aluminium bronze made by Thomas Shaw, Newark, N. J.; Cowles Bros.' reports of the strength of aluminium silver castings; Solders for aluminium bronze for jeweller's use	279
Silicon and aluminium bronze, Cowles Bros.; Aluminium and iron; Tissier Bros.' alloy; Deville, Rogers	280
Fremy and Mierzinski on aluminium alloys	282
Ostberg's mitis castings, with description of the process	283
Alloy used by Ostberg, Worcester, Mass.	285
Ostberg's note to the Engineering and Mining Journal; Watts's note; Mr. Sellers's series of experiments	286
Aluminium and zinc; Tissier Bros., Deville, Kerl and Stohman, and Fremy on these alloys	287
Aluminium and tin; Tissier Bros., Deville, and Kerl and Stohman on this subject	289
Fremy, Mierzinski, and M. Bourbouze	290
Aluminium and lead; Tissier, Deville, Kerl and Stohman, and Mierzinski	291
Aluminium and antimony; Tissier, and Kerl and Stohman; Aluminium and bismuth; Tissier and Watts	292
Aluminium and nickel; Tissier and Mierzinski	293
Argentan	294
Minargent	295
Aluminium and silver; Tissier on this subject	295

CONTENTS. xix

	PAGE
Deville; Kerl and Stohman; Fremy	296
Mierzinski; "Tiers Argent;" Cowles Bros. on "Aluminium silver"	297
Aluminium and gold; Tissier, Fremy, and Mierzinski on these alloys	298
Aluminium and platinum; Tissier; Aluminium and Cadmium, Deville; Aluminium and boron, Deville on this subject	299
Aluminium and carbon, Deville and Cowles; Aluminium and gallium, Watts, Lecoq de Boisbaudran	300
Aluminium and titanium, Wöhler	301
Aluminium and tungsten, by Michel, in Wöhler's laboratory; Aluminium and molybdenum; Experiments by Michel; Aluminium and manganese; Experiments by Michel	302
Aluminium and sodium; Deville and Fremy on these alloys; Aluminium and nitrogen, Dr. Hunt	303

APPENDIX.

Native sulphate of alumina; Account of a deposit in New Mexico 305
Decomposition of cryolite; F. Lauterborn's patent; American aluminium; Frishmuth's metal . . . 306
Analyses of same; Specific gravity of aluminium; Gravity calculated from analyses 307
Amalgamation of aluminium 308
Theory of the rapid oxidation of aluminium amalgam; Reduction of alumina; Experiment on reduction with copper; Production and reduction of aluminium sulphide; Fremy's researches 309
Investigations by Reichel 312
Than's remarks 314
Reichel's experiments in reducing aluminium sulphide; Petitjean's patent 315

CONTENTS.

	PAGE
Reichel's summary	316
Aluminium chloride formed from the sulphide; F. Lauterborn's patent	317
Niewerth; Reichel; A. Orlowski; Experiments on making aluminium sulphide	318
Tabulated results; Remarks and suggestions of a practical process	321
Reducing the aluminium sulphide	322
Experiments with lead, copper, tin, antimony, and iron; Review of these experiments and suggestions of a practical process	323

ADDENDA.

Additional details of Castner's sodium process	324
New process for making aluminium chloride; Remarks on the mitis process	326
Production of aluminium in 1885; Low price of aluminium in October, 1886	327
INDEX	328

ALUMINIUM.

PART I.

HISTORY OF ALUMINIUM.

LAVOISIER* first suggested the existence of metallic bases of the earths and alkalies. The first researches in the preparation of aluminium date back to 1807. Davy tried, but in vain, to decompose Al^2O^3 by an electric current, or to reduce it by vapor of potassium. Oerstedt, in 1824, believed he had isolated aluminium. He decomposed anhydrous Al^2Cl^6 by K amalgam, and he obtained, along with some KCl, an amalgam which decomposed by heat furnished him a metal resembling tin. It is probable that he employed either some moist Al^2Cl^6 or K amalgam which contained KOH, for it is only when wetted with a solution of KOH that aluminium alloys with mercury; for, when Wöhler, later, wished to prepare aluminium by this method, he found it impossible to obtain an Al amalgam when he employed materials pure

* Fremy, Ency.

and dry. Nevertheless, the method of Oersted marks an epoch in the history of the science, for in 1827 Wöhler isolated aluminium by decomposing Al^2Cl^6 by K. The metal first isolated by Wöhler was a gray powder, taking under the polisher the brilliancy of tin. It was very easily changed, because of its extreme division, and also because it was mixed with the K or Al^2Cl^6 used in excess. At that time no further use was made of these facts. Later, in 1845, on making vapor of Al^2Cl^6 pass over potassium placed in platinum boats, Wöhler obtained the metal in small, malleable globules of metallic appearance, from which he was able to determine the principal properties of aluminium. But the metal thus obtained was scarcely as fusible as cast iron, without doubt because of the platinum with which it had alloyed during its preparation. In addition to this, it decomposed water at 100°, from which we suppose that it was still impregnated with K or Al^2Cl^6. It is to H. St. Claire Deville that the honor belongs of having in 1854 isolated aluminium in a state of almost perfect purity, determining its true properties. In commencing researches on aluminium, Deville, while he applied the method of Wöhler, was ignorant of the latter's results of 1845. Besides, he was not seeking to produce aluminium that he might turn its valuable properties to practical account, but that it might serve for the production of AlO, which he believed could exist as well as

FeO. The aluminium he wished to prepare would, he thought, by its further reaction on Al^2Cl^6 form $AlCl^2$, from which he might derive AlO and the other proto-salts. But this proto-chloride was not thus produced; he obtained, enclosed in a mass of $Al^2Cl^6.2KCl$, fine globules of a brilliant substance, ductile, malleable, and very light, capable of being melted in a muffle without oxidizing, attacked by HNO^3 with difficulty, but dissolved easily by HCl or KOH with evolution of hydrogen. Recognizing the importance of these properties, which he proceeded to investigate and establish, and fearing to see the honor of his discovery pass into other hands, Deville immediately commenced research for an economic process to produce aluminium. The task was difficult, for the metal could only be isolated from its chloride or fluoride by potassium or sodium, of which only the former was known at that time. Moreover, potassium cost then 900 fr. per kilo, was extremely dangerous, and gave only a small return of aluminium. Deville succeeded in advantageously replacing potassium by sodium, and introduced such improvements into the manufacture of the latter that he reduced the cost of a kilo from 2000 fr. in 1855 to 10 fr. in 1859. In order to produce aluminium cheaply, he busied himself also in the economic production of Al^2O^3, which gave later a lively impulse to the cryolite and beauxite industries. The researches of Deville, at first undertaken in the laboratory of the Normal School, Paris, were afterwards continued on a larger scale, thanks to the liberality of the Emperor Napoleon III., at the chemical works of Javel. At this works were made the ingots and divers objects of aluminium which figured at the Paris Exhibition in 1855. Later, after new experiments made together at the Normal School, Deville, H. Debray, and P. Morin set up a plant to make aluminium on a large scale at Messrs. Rousseau Brothers' works at Glacière. The primary method there received many improvements. Later on it was still further improved under the direction of P. Morin at the works in Nanterre. At last, in the works of Messrs. Merle & Co., at Salindres, it has reached its present stage of advancement.

Tissier Bros. wrote and published a book entitled 'Recherches sur l'Aluminium' in 1858. These brothers were employed in the experiments which Deville superintended at the laboratory of the Normal School, Paris, and Deville charges that after learning the important results of his experiments they suddenly left him, taking drawings of furnaces, details of processes, etc., and started works themselves. Deville was very bitter against them, and this ill-feeling was increased by the following incident: Deville was collecting material to write a book on the subject, which he almost regarded as his prerogative, seeing that he had, so to speak, created the industry; but, before he had completed it, Tissier Bros. published theirs. In order not to be too far behind, Deville hastened the completion of his book, by doing which he was unable to make it as full as he had wished, and published it in September, 1859. Several sharp letters passed between Deville and the Tissiers, which may be seen in the Compt. Rend. or Ann. de Chem. et de Phys. A. Tissier, in his book, thus describes the formation and history of his works: In July, 1855, Messrs. Malétra, Chanu, and Davey, of Rouen, formed a company to produce aluminium, and we were entrusted with the organization and special charge of the industry. The commencement was beset with difficulties, not only in producing, but in using the metal. It then sold at $200 per kilo, the price being an insurmountable obstacle to its employment in the arts. The small capital at our disposal was not enough to start the industry, to pay general expenses, and the losses occasioned by the many experiments necessary. On February 28, 1856, the society was dissolved. In April, the same year, Mr. William Martin, struck by the results already obtained, and sanguine of greater success, united with us. From that time daily improvements confirmed M. Martin's hopes, and in 1857 the works at Amfreville-la-mi-Voie, near Rouen, sold the metal at $60 per kilo ($2.00 per oz.). The laboratory of this works was devoted to researches on everything concerning the production and application of aluminium. M. Martin has our sincere gratitude for the kindness with which he so willingly encouraged and contributed to the progress of the manufacture of "this wonderful metal."

Deville, as stated above, published his book in September, 1859, and he concludes it with these words: "I have tried to show that aluminium may become a useful metal by studying with care its physical and chemical properties, and showing the actual state of its manufacture. As to the place which it may occupy in our daily life, that will depend on the public's estimation of it and its commercial price. The introduction of a new metal into the usages of man's life is an operation of extreme difficulty. At first, aluminium was spoken of too highly in some publications, which made it out to be a precious metal; but later these estimates have depreciated even to the point of considering it attackable by pure water. The cause of this is the desire which many have to see taken out of common field mud a metal superior to silver itself; the opposite opinion established itself because of very impure specimens of the metal which were put in circulation. It seems now that the intermediate opinion, that which I have always held and which I express in the first lines of my book, is becoming more public, and will stop the illusions and exaggerated beliefs which can only be prejudicial to the adoption of aluminium as a useful metal. Moreover, the industry, established as it now is, can be the cause of loss to no one; as for myself, I take no account of the large part of my estate which I have devoted, but am only too happy, if my efforts are crowned with definite success, in having made fruitful the work of a man whom I am pleased to call my friend—the illustrious Wöhler."

As early as 1856 we find an article in an American magazine* showing that there were already chemists in the United States spending time and money on this subject. The following is the substance of the article alluded to: "Within the last two years Deville has extracted 50 to 60 lbs. of aluminium. At the present time, M. Rousseau, the successor of Deville in this manufacture, produces aluminium which he sells at $100 per pound. No one in the United States has undertaken to make the metal until recently Mons. Alfred Monnier, of Camden, N. J., has, according to the statement of Prof. James C. Booth in the 'Penn. Inquirer,' been successful in making sodium by a continuous process, so as to procure it in large bars, and has made aluminium in considerable quantity, specimens of which he has exhibited to the Franklin Institute. Mons. Monnier is desirous of forming a company for the manufacture of aluminium, and is confident that by operating in a large way he can produce it at a much less cost than has hereto-

* Mining Magazine, 1856, vii. 317.

fore been realized. We would suggest the propriety of giving aid to this manufacturer at the expense of the government, for the introduction of a new metal into the arts is a matter of national importance, and no one can yet realize the various and innumerable uses to which this new metal may be applied. It would be quite proper and constitutional for Congress to appropriate a sum of money, to be expended under the direction of the Secretary of the Treasury in the improvement of this branch of metallurgy, and in testing the value of the metal for coinage and other public use."

In the next volume of the 'Mining Magazine'[*] there is a long article by Mr. W. J. Taylor, containing nothing new in regard to the metallurgy of aluminium, but chiefly concerned in calculating theoretically the cost of the metal from the raw materials and labor required by Deville's processes, and concluding that it is quite possible to make it for $1.00 per pound.

In 1874 we have the following *résumé* by Kerl & Stohman: " Deville first worked under the direction of the Paris Academy; later, the Emperor Napoleon gave him great encouragement, by means of which he succeeded in producing several kilos of aluminium, which were shown at the exhibition in Paris, 1855. With the experience thus gained, Deville took possession of Rousseau Bros.' chemi-

[*] Mining Magazine, viii. 167 and 228. Proc. Ac. Nat. Sci., Jan. 1857.

cal works at La Glacière, near Paris. It soon followed that the price of aluminium was reduced from 1000 fr. per kilo to 300 fr. After a short time the undertaking was enlarged, and the manufacture removed to Nanterre and Salindres. The last named works, under the management of Usiglio, went into the possession of Merle. New advances made a further reduction in price to 200 fr. possible. In 1862 the price was put down to 130 fr. Another works was then establishèd at Amfreville, near Rouen. This was on a larger scale than that at Nanterre, for while in 1859 the latter produced 60 kilos, the former produced 80. In England the first manufactory was established in 1859, at Battersea, London; and the next year Bell Bros. started at Newcastle-on-Tyne. Germany as yet possesses no aluminium works."

The further we get away from an age the better able are we to write the true history of that age. And so, as years pass since the labors of Wöhler, Deville, and Tissier, we are now able to see better the whole connected history of the development of this art. Dr. Clemens Winckler gives us a comprehensive retrospect of the field seen from the standpoint of 1879, from which we condense the following:* The history of the art of working in aluminium is a very short one, so short that the present generation, with which it is contemporary, is in

* Industrie Blätter, 1879; Sci. Am. Suppl., Sept. 6, 1879.

danger of overlooking it altogether. The three international exhibitions which have been held in Paris since aluminium first began to be made on a commercial scale form so many memorials of its career, giving, as they did, at almost equal intervals, evidence of the progress made in its application. In 1855, we meet for the first time, in the Palais de l'Industrie, with a large bar of the wonderful metal, docketed with the extravagant name of the "silver from clay." In 1867 we meet with it again, worked up in various forms, and get a view of the many difficulties which had to be overcome in producing it on a large scale, purifying, and moulding it. We find it present as sheets, wire, foil, or worked-up goods, polished, engraved, and soldered, and see for the first time its most important alloy—aluminium bronze. After a lapse of almost another dozen years we see at the Paris exhibition of 1878 the maturity of the industry. We have passed out of the epoch in which the metal was worked up in single specimens, showing only the future capabilities of the metal, and we see it accepted as a current manufacture, having a regular supply and demand and being in some regards commercially complete. The despair which has been indulged in as to the future of the metal is thus seen to have been premature. The manufacture of aluminium and goods made of it has certainly not taken the extension at first hoped for in its behalf; the lowest limit of the cost of manufacture was soon reached, and aluminium remains as a metal won by expensive operations from the cheapest of raw materials.

To France is due the merit of having been the first country to carry out Wöhler's process on a practical scale, and to have created the aluminium industry. France seems to be the only country in which the industry is able to prosper. The English establishment at Newcastle-on-Tyne by Bell & Co. did not succeed, and has been shut up now for about five years. The German manufactory, set up in Berlin by Wirz & Co., cannot be said really to have lived at all; it drooped before it was well started. In France, the great chemical works of H. Merle & Co., Salindres, carries on the extraction of aluminium, and the Société Anonyme de l'Aluminium, at Nanterre, works up the metal. Both firms were represented at the exhibition in 1878.

The most rational use indicated for aluminium by reason of its low specific gravity is the making of beams for balances. Sartorius, of Göttingen, was the first to make these light and unalterable beams of an alloy of 96 aluminium and 4 silver. He has had but few imitators. There are several reasons why the metal is shown so little favor by mathematical instrument makers and others. First of all, there is the price; then the methods of working it are not everywhere known; and further, no one knows how to cast it. Molten aluminium attacks the common earthen crucible, reduces silicon from it, and becomes gray and brittle. This inconvenience is overcome by using lime crucibles, or by lining an earthen crucible with carbon or strongly burnt cryolite clay. If any one would take up the casting of aluminium and bring it into vogue as a current industrial operation, there is no doubt that the metal would be more freely used in the finer branches of practical mechanics. The prices per kilo quoted in the last list issued by the Société Anonyme are as follows:—

ALUMINIUM.

Bars	130 fr.
Sheets (0.5 to 0.1 mm. thick)	135 " to 160 fr.
Wire (2 to 3 mm. diam.)	170 " " 200 "

ALUMINIUM BRONZE (10 per cent. aluminium).

Bars	18 fr.
Sheets (2 to 0.5 mm. thick)	24 " to 30 fr.
Wire (7 to 1 mm. diam.)	28 " to 39 "

The preceding paper of Dr. Winckler, as he remarks, chronicles the perfection of Deville's processes, when aluminium was made as cheaply as it could possibly be by these methods. But, about this time an aluminium works was started in Birmingham, England, by Mr. Webster, which has grown to be one of the largest in the world. Mr. Webster owns several patents on processes of his own, which will be found described in their proper places.

In the United States one of the most prominent

chemists engaged on aluminium is Colonel William Frishmuth, of Philadelphia. The following article gives an account of his invention :* " Some months ago, we published in the 'Star' the fact that Colonel William Frishmuth, well known in this city for many years, has discovered a method for producing aluminium at reduced cost. Comments were made in various quarters as to the real value of the discovery, some of which even questioned the possibility of producing the metal by this process, which is stated to produce it from South Carolina corundum, using sodium as a reagent. Meanwhile patents have been taken out in this and foreign countries, and preliminaries are fairly under way to test the process practically. It did not seem too much to hope when the publication was made that American capitalists would at once make investigation of Colonel Frishmuth's discovery, learn whether the results were even measurably up to the promise, and in that event secure to themselves a commercial plant so extremely important. It has, however, fallen to capitalists abroad to obtain control of the patent. At the present time Major Ricarde-Seaver, F.R.S.E., late Government Inspector of Mines, London, is in this city as an expert to examine the process and its practicability on behalf of these capitalists. A reporter endeavored to obtain from Major Seaver his opinion of the process, but he

* Philadelphia Evening Star, November 15, 1884.

stated that his opinion could not be made public. Mr. Seaver said in reference to aluminium: 'Some of the best minds in Europe have been studying for years the problem of producing the metal cheaply. Scientists in France and Germany and one in Geneva have been at work on it a long time. As to the possibility of producing it so that it could be used as an alloy for iron and steel, that is not to be expected unless it could be produced at much less than a dollar per pound. As to the possibility of doing that by this process, I am not at liberty to speak. The work here has so far been merely on an experimental scale. As scientific men know by many experiences, disappointments are sometimes met with when they leave the experimental field and work is attempted on a commercial scale for business purposes. I am, however, very much pleased with what I have seen here, and, as I have said, while scientific men all over Europe have been investigating the problem, it seems to be solved here. I am certainly satisfied that aluminium can be produced by Frishmuth's process, there is some metal made by it, and there will be a display of it at the New Orleans exhibition. Even if it can be made very cheap by this process, it is not probable that anything more would be done by the parties I represent than to supply the market at a fair price, just as the Rothschilds, who own the great quicksilver mines of the world, regulate the supply by the demand.'"

Colonel Frishmuth's works are at Rush and Amber streets, near the Richmond coal wharves, Philadelphia. He seems to be kept busy, and his metal is on the market; an analysis of it will be found in the Appendix. It can be bought from Bullock & Crenshaw, Philadelphia. He cast the tip of the Washington Monument, which weighs one hundred ounces, one of the largest single castings of aluminium ever made. As far as is known, he is at present the only producer of pure aluminium in the United States. His metal is sold in bars at about fifteen dollars per pound. In the Philadelphia city census of 1884 he is placed as employing ten men, and his annual product is valued at $18,000. Mr. Frishmuth melts down quantities of aluminium scrap, and the author has been unable to learn, except from Mr. Seaver's statement, that Mr. Frishmuth produces any aluminium by his process. Mr. Seaver represented an English syndicate which stands ready to buy out all patents of any value which appear on aluminium; they possess large capital, and are said to be ready to pay an immense sum for any practical, cheap process for producing the metal.

In the Mineral Resources of the United States, 1883–4, we find a few statistics as to the amount of aluminium made in recent years. It is there stated that in 1882 there were 2350 kilos made in France. The price of the American metal ranged from $0.75 to $1.00 per troy ounce in 1883; and from $0.50 to $1.00 per ounce in 1884, according to quantity. The amount imported and entered for consumption in the United States from 1870 to 1884 is as follows:—

Year ending June 30,	Quantity (pounds).	Value.
1870	...	$ 98
1871	...	341
1872		
1873	2	22
1874	683	2125
1875	434	1355
1876	139	1412
1877	131	1551
1878	251	2978
1879	284	3423
1880	341	4042
1881	517	6071
1882	566	6459
1883	436	5079
1884	590	8416

Until recently the aluminiun sold in the United States was entirely of foreign origin, but it is now produced in this country by Colonel Frishmuth, of Philadelphia, who turned out 1000 ounces of the metal in 1883, and 1800 ounces in 1884. The aluminium cap or apex of the Washington Monument was cast by him; it is of pyramidal form, 10 inches high, 6 inches on a side of its base, and weighs 8½ pounds (see p. 53).

Within the last two years a process has been invented and brought into practical use which has served to bring the metallurgy of aluminium into

HISTORY OF ALUMINIUM. 41

very general attention. The Cowles' process, the discovery and details of which will be given further on, is due to two Cleveland gentlemen, and they seem to be developing all that is in their process. They make no pure metal, but sell the alloys, principally aluminium bronze, the latter of good quality, and at a much lower price than it was ever sold before. If they can make it profitable to sell the bronze at the price which they now quote, the permanent success of their process is assured. Mr. Charles F. Mabery, of the Case School of Applied Science, Cleveland, is their consulting chemist, and Dr. T. Sterry Hunt, of Montreal, seems to be very much interested in the process from a scientific point of view. Mr. Mabery gives his views as to the present state of the aluminium industry as follows: "The aluminium of commerce has been made chiefly in France by Deville's old method. Several patents have been issued for its production by electrolysis, and although it can be deposited in small quantities from solutions, there is but one electrolytic method that can be worked on a commercial basis, and that is Bunsen & Deville's method of electrolysing molten $Al^2Cl^6.2NaCl$. Large works have recently been erected in France for obtaining the metal by this method, and it is claimed that it can be produced for about \$7 per pound. A company has recently been formed in London to manufacture aluminium alloys on the basis of the Webster patents. The chief improvement on the old process, according to the patent specifications, is in the preparation of the pure Al^2O^3. Frishmuth, of Philadelphia, attempts to produce sodium in one retort, volatilize aluminium chloride from another, and allow the vapors to meet in a third. The assertion made by him at first that he could place the metal on the market at $1.25 per pound has not been verified." Dr. Hunt, in reply to an inquiry as to the present state of the industry, replies: " Webster, of England, is the chief, perhaps the only, manufacturer in that country of the metal and its alloys. Messrs. Cowles manufacture the alloys, and they can now make pure aluminium, but the method is not yet perfected or made public. The process of Frishmuth is not new, but is mentioned in Watts' Dictionary. So far as I can learn, and so far as Messrs. Cowles are informed, there has been no pure aluminium made commercially save from the chloride by use of sodium. Messrs. Cowles' work with their large new dynamo has been very satisfactory."

PART II.

OCCURRENCE OF ALUMINIUM IN NATURE.

THERE is no other metal on the earth which is so widely scattered and occurs in such abundance. Al is not found metallic. Stocker* made the statement that Al occurred as shining scales in an alumina formation at St. Austel, near Cornwall, but he was in error. But the combinations of Al with oxygen, the alkalies, fluorine, silicon, and the acids, etc., are so numerous and occur so abundantly as not only to form mountain masses, but to be also the bases of soils and clays. Especially numerous are the combinations with Si and other bases, which, in the form of felspar and mica, mixed with quartz, form granite. Mierzinski gives the formulæ of a few of these silicates as:—

Orthoclase	$K^2Si^3O^7 + Al^2Si^3O^9$
Albite	$Na^2Si^3O^7 + Al^2Si^3O^9$
Anorthite	$CaSiO^3 + Al^2SiO^5$
K Mica	$(HK)^2Al^2Si^2O^8 + (HK)^2Al^2Si^4O^{15}$
Na Mica	$(HNa)^2Al^2Si^2O^8 + (HNa)^2Al^2Si^4O^{15}$
Li Mica	$(Li^2.K^2.Fe)O.Al^2O^3.3SiO^2.$
Mg Mica	$m(HK)^4SiO^4 + n(Mg.Fe.H.)^2SiO^4 +$ $p(Al^2Fe^2)^2Si^3O^{12}$

---

* Jrnl. fr. prakt. Chem., 66, 470.

These combinations, by the influence of the atmosphere, air, and water, are decomposed, the alkali is replaced or carried away, and the residues form clays. The clays form soils, and thus the surface of the earth becomes porous to water and fruitful. It is a curious fact that Al has never been found in animals or plants, which would seem to show that it is not necessary to their growth, and perhaps would act injuriously, if it were present, by its influence on the other materials. Most of the Al compounds appear dull and disagreeable, such as felspar, mica, pigments, gneiss, amphibole, porphyry, eurite, trachyte, etc.; yet there are others possessing extraordinary lustre, and so beautiful as to be classed as precious stones. Some of these, with their formulæ, are—

Ruby	Al^2O^3
Sapphire	Al^2O^3
Garnet	$(Ca.Mg.Fe.Mn)^3 Al^2 Si^3 O^{12}$
Cyanite	Al^2SiO^5

Some other compounds occurring frequently are—

Turquoise	$Al^2P^2O^8.H^6Al^2O^6.2H^2O$
Lazulite	$(MgFe)Al^2P^2O^9 + Aq$
Wavellite	$2Al^2P^2O^8.H^6Al^2O^5.9H^2O$
Topaz	$5Al^2SiO^5.Al^2SiF^{10}$
Cryolite	$Al^2F^6.6NaF$
Diaspore	$H^2Al^2O^4$
Beauxite	$H^6Al^2O^6$
Aluminite	$Al^2SO^6.9H^2O$
Alunite	$K^2SO^4.Al^2S^3O^{12}.2H^2Al^2O^3$

One would suppose that since aluminium occurs in such abundance over the whole earth, since we literally tread it under foot, that it would be extracted and applied to numberless uses, being made as abundant and useful as iron; but such is not the case.

Beauxite and cryolite are the minerals most used for producing aluminium, and their preference lies mainly in their purity. Native alums generally contain Fe, which must be removed by expensive processes. Some observations on a native alum deposited in New Mexico will be found in the Appendix. We will here consider at greater length only beauxite and cryolite.

BEAUXITE.

Beauxite is a combination between diaspor, $Al^2O^3.3H^2O$, and brown hematite, $Fe^2O^3.3H^2O$; or, it is diaspor with Al replaced more or less by Fe; the larger the amount of Fe the more its color changes from white to brown. It was first found in France, near the town of Beaux, large deposits occurring in the departments Var and Bouches du Rhon, extending from Tarascon to Antibes. Several of these beds are a dozen yards thick, and 160 kilometres in length. Deposits are also found in the departments of l'Herault and l'Arriège. Very important beds are found in Styria, at Wochein, and at Freibriss, in Austria, a newly discovered locality where the mineral is called Wocheinite. Here it has a dense, earthy structure, while that of France is conglomerate or oölitic. Deposits similar to those of France are found in Ireland at Irish Hill, Straid, and Glenravel. Further deposits are found in Hadamar in Hesse, at Klein-Steinheim, Langsdorff, and in French Guiana.

The following analyses give an idea of the peculiar composition of this mineral; besides the ingredients given there are also traces of CaO, MgO, SO^3, P^2O^5, TiO^2, and Va^2O^3.

	a.	b.	c.	d.	e.	f.
Al^2O^3	60.0	75.0	63.16	72.87	44.4	54.1
Fe^2O^3	25.0	12.0	23.55	13.49	30.3	10.4
SiO^2	3.0	1.0	4.15	4.25	15.0	12.0
K^2O and Na^2O	...	...	0.79	0.78	...	...
H^2O	12.0	12.0	8.34	8.50	9.7	29.9

	g.	h.	i.	k.	l.	m.
Al^2O^3	64.6	29.80	48.12	43.44	61.89	45.76
Fe^2O^3	2 0	3.67	2.36	2.11	1.96	18.96
SiO^2	7.5	44.76	7.95	15.05	6.01	6.41
K^2O and Na^2O	...	...	...	...	...	0.38
H^2O	24.7	13.86	40.33	35.70	27.82	27.61

	n.	o.	p.	q.	r.
Al^2O^3	55.61	76.3	50.85	49.02	73.00
Fe^2O^3	7.17	6.2	14.36	12.90	4.26
SiO^2	4.41	11.0	5.14	10.27	2.15
K^2O and Na^2O	...	...	0.26	0.31	...
H^2O	32.33	26.4	28.38	25.91	18.66

OCCURRENCE OF ALUMINIUM IN NATURE. 47

Index:—

a and b. from Beaux (Deville).
c.   dark ⎫ Wocheinite (Drechsler).
d.   light ⎭
e.   red brown ⎫
f.   yellow   ⎬ Beauxite from Feisstritz (Schnitzer).
g.   white   ⎭
h.   white Wocheinite (L. Mayer and O. Wagner).
i.   Beauxite from Irish Hill.
k.   "   "   Co. Antrine (Spruce).
l.   "   "   Glenravel (F. Hodges).
m and n. "   "   Hadamar (Hesse) (Retzlaff).
o.   from Klein-Steinheim (Bischof).
p and q. from Langsdorff (I. Lang).
r.   Beauxite from Dublin, Ireland, brought to the Laurel Hill Chemical Works, Brooklyn, L. I., and there used for making alums. It is dirty white, hard, dense, compact, and in addition to the ingredients given above contains 0.59 per cent. CaO, and some TiO^2. It costs $6 per ton laid down in the works. The above analysis, made by Mr. Jouet, is furnished me by the kindness of the superintendent of the works, Mr. Herreshoff.

As is seen from the above analyses, the percentage of Al^{2}O^3 is very variable, and cannot be determined at all simply by inspection but only by an analysis, for often the best-looking specimens are the lowest in Al^{2}O^3. For instance, a beauxite containing 62.10 Al^{2}O^3, 6.11 Fe^{2}O^3, 5.06 SiO^2, and 20.83 H^{2}O was much darker and more impure looking than that from Wochein (h) which contained only 29.8 per cent. Al^{2}O^3.

Cryolite.

Cryolite was first found at Ivigtuk, in Arksutfiord, west coast of Greenland, where it constitutes a large bed or vein in gneiss. It is a semi-transparent, snow-white mineral. When impure it is yellowish or reddish, even sometimes almost black. It is shining, sp. gr. 2.95, and hardness 2.5 to 3. It is brittle, not infrequently contains $FeCO^3$, PbS, SiO^2, and sometimes columbite. It is fusible in the flame of a candle, and on treatment with sulphuric acid yields hydrofluoric acid. As will be seen further on, cryolite was first used by the soapmakers for its soda; it is still used for making soda and alumina salts, and to make a white glass which is a very good imitation of porcelain. The Pennsylvania Salt Company in Philadelphia import it from Ivigtuk by the ship-load for these purposes; lately they have discontinued making the glass. Cryolite is in general use as a flux. A very complete description of the deposit at Ivigtuk can be found in Hoffman's 'Chemische Industrie.'

The only known deposit of cryolite in the United States is that found near Pike's Peak, Colorado, and described by W. Cross and W. F. Hillebrand in the 'American Journal of Science,' October, 1883. It is purely of mineralogical importance and interest, occurring in small masses as a subordinate constituent in certain quartz and feldspar veins in a country rock of coarse reddish granite. Zircon, astrophyllite, and columbite are the primary associated minerals, the first only being abundant.

There is no duty on the imports of cryolite into the United States, and they have varied from 10,000 tons in 1869 to 9000 in 1884, costing $9 to $10 per ton.

CORUNDUM.

"Till 1869, the sole sources of corundum were a few river washings in India and elsewhere. It was found in scattered crystals, and cost twelve to twenty-five cents per pound. In 1869, in riding over a spur of the Alleghenies in northern Georgia, I* found what has proven to be an almost inexhaustible mine of corundum in the crysolite serpentine, the first instance on record of the mineral being found *in situ*. Previously it had been washed out of débris at Cripp's Hill, N. C., and at a mine in West Chester, Pa., both on the slopes of the crysolite serpentine. The clue being thus obtained accidentally, about thirty mines were shortly afterwards discovered in the same formation; but of the thousands of tons thus far dug out, the larger portion has come from the mines I discovered.

"At present it can be bought at about ten dollars per ton at the mines. It is nearly pure Al^2O^3.

(* Mr. W. P. Thompson.

Disapore, a hydrated alumina, is also found in the same region and locality. Corundum will probably always be the principal source in America of material from which to manufacture pure Al. But in Great Britain, in all probability, manufacturers must look to alumina prepared artificially from cryolite or from Mr. Kynaston's sulphate of alumina."*

* Journal of the Society of Chemical Industry, April, 1886.

PART III.

PHYSICAL PROPERTIES OF ALUMINIUM.

COMMERCIAL aluminium is never chemically pure, and therefore displays properties varying more or less from those of the pure metal according to the character and amount of impurities present. In this treatise, whenever the properties of aluminium are mentioned, they must be understood to refer to the chemically pure metal, and not to the commercial article, unless specifically stated. As preliminary to the presentation of these properties we will here make some observations on the commercial metal and the impurities generally found in it.

In whatever way aluminium may be reduced, still it is always far from being pure, being contaminated with iron, silicon, or even sodium and lead, as is shown by the following analyses:—

	$Al.$	$Si.$	$Fe.$	$Cu.$	$Pb.$	$Na.$
1. Parisian (Salvêtat)	92.969	2.149	4.882	..	..	trace
2. Berlin (Mallet)	96.253	0.454	3.293	..	..	trace
3.	96.890	1.270	1.840	trace	..	..
4. Morin & Co., Nanterre (Sauerwein)	97.200	0.25	2.40	..	trace	trace
5. Parisian (Dumas)	92.500	0.70	6.80	..	..	..
6.	96.160	0.47	3.37	..	..	..
7. Parisian (Salvêtat)	88.350	2.87	2.40	6.38	trace	..
8. Parisian, from Haag	92.000	0.45	7.55	..	..	..
9. Bonn (Kraut)	94.700	3.70	1.60	..	..	..
10. Morin & Co., Nanterre, 1862 (Kraut)	98.290	0.04	1.67	..	..	..
11.	97.680	0.12	2.20	..	..	..
12. (Hampe) the purest he could buy	97.400	1.00	1.30	0.10	0.20	..
13. Wagner's Jahresb., 1877	97.600	0.40	1.40	0.40	0.20	..

According to Rammelsberg (Kerl's Handbuch) the Si which is always found in aluminium is in part combined with it, and this combined Si changes by treatment with HCl into either SiO^2, which remains, or into SiH^4, which escapes; while another part of the Si is combined with the aluminium just as graphite is with Fe; and this part of the Si remains on treatment with acid as a black mass, not oxidized by ignition in the air. Two analyses of aluminium reduced from cryolite by sodium in a porcelain crucible gave—

	1.	2.
SiO^2	9.55	1.85
Free Si	0.17	0.12
Si in SiH^4	0.74	0.58

One sample of aluminium analyzed by Professor Rammelsburg contained as much as 10.46 per cent. Si, and another sample even 13.9 per cent. The quantity of Fe varies from 2.9 to 7.5 per cent.

M. Dumas has found that aluminium usually contains gases, about which he makes the following statements:* "On submitting aluminium in a vacuum to the action of a gradually increasing temperature up to the softening point of porcelain, and letting the mercury pump continue acting on the retort until it was completely exhausted, considerable quantities of gas were withdrawn. The liberation of the gas from the metal seems to take

---

* Sci. Am. Suppl., Aug. 7, 1880.

place suddenly towards a red-white heat. 200 grammes of aluminium, occupying 80 c. c., gave 89.5 c. c. of gas, measured at 17° and 755 mm. pressure. The gas consisted of 1.5 c. c. CO^2 and 88 c. c. H. CO, N, and O were absent."

*The aluminium apex or cap of the Washington Monument cast by Colonel Frishmuth, of Philadelphia, has the following composition:—

Al	97.75
Fe	1.70
Si	0.55

COLOR.

Deville: The color of aluminium is a beautiful white with a slight blue tint, especially when it has been strongly worked. Being put alongside silver, their color is sensibly the same. However, common silver, and especially that alloyed with copper, has a yellow tinge, making the aluminium look whiter by comparison. Tin is still yellower than silver, so that aluminium possesses a color unlike any other useful metal.

Fremy: Aluminium has a fine white color, just a little blue when compared with silver. When it has been worked, or when it contains Fe or Si, its blue tint acquires greater intensity. The commercial aluminium resembles silver.

* Mineral Resources of the United States, 1883–84.

Mallet: Absolutely pure aluminium is perceptibly whiter than the commercial metal; on a cut surface very nearly pure tin-white, without bluish tinge, as far as could be judged from the small pieces examined.

Mierzinski: The pure white color of aluminium is very brilliant; it has a tint lying between the color of tin and zinc, although on account of its usual blue shading, even in a poor light, it cannot be confounded with them or with any white metal.

Mat.

Deville: Aluminium like silver is able to take a very beautiful mat which keeps indefinitely in the air. It is obtained easily by plunging the surface for an instant in a very dilute solution of caustic soda, washing in a large quantity of water and at last dipping in strong nitric acid. Under these conditions, all the foreign materials which might contaminate it, except silicon in large proportion, dissolve and leave the metal quite white and with a very pleasing appearance.

Mierzinski: The peculiar lustre of aluminium, however, is not permanent. With time, the objects take on their plain faces an olive green coloration, and look much less agreeably. Their former white color can be restored by Mourey's receipt, by placing them first in dilute hydrofluoric acid, 1000 parts water to 2 of acid, and then dipping them in nitric acid.

Bell Bros.: They recommend first washing the objects in benzole or essence of turpentine, before treating with NaOH and HNO^3, as above.

Polish and Lustre.

Deville: Aluminium may be polished and burnished easily, but it is necessary to employ as an intermediate material between the stone and polishing powder a mixture of stearic acid and essence of turpentine, finishing with pure essence of turpentine. In general, the polished surfaces are of a less agreeable appearance than the mat, the blue tint of the metal becoming more manifest. But, in this work, the experience and practice of the workers in aluminium is far from being complete; each metal requires a special way of working, and we may expect yet for a material so new that progress will be made in this direction.

Bell Bros.: Aluminium is easily polished and burnished. Use a mixture of equal parts of rum and olive oil as an intermediate substance between the polishing stone and the powder used. The polishing stone is steeped in this mixture, and will then burnish the metal as silver and copper are burnished, care being taken not to press too heavily on the burnishing instrument.

Kerl & Stohman: The use of the old means of

polishing and burnishing metals, such as soap, wine, vinegar, linseed-oil, decoction of marshmallow, etc., is not effective with aluminium, but, on the contrary, is even harmful; because, using them, the blood stone and the burnishing iron tear the metal as fine stone does glass. Oil of turpentine has also been used, but with no good effect. Mourey found, after many attempts, that a mixture of equal weights of olive oil and rum, which were shaken in a bottle till an emulsified mass resulted, gave a very brilliant polish. The polishing stone is dipped in this liquid, and the metal polished like silver, except that one must not press so hard in shining up. The peculiar black streaks which form under the polishing stone need cause no trouble; they do not injure the polish in the least, and can be removed from time to time by wiping with a lump of cotton. The best way to clean a soiled surface and remove grease is to dip the object in benzine, and dry it in fine sawdust. Hammered and pressed objects of aluminium may, before polishing, be very easily ground by using olive oil and pumice.

Odor.

Deville: The odor of pure aluminium is sensibly nothing, but the metal strongly charged with silicon will exhale the odor of silicuretted hydrogen, exactly represented by the odor of cast iron. But, even under these unfavorable circumstances, the smell of the metal is only appreciable to persons experienced in judging very slight sensations of this kind.

Watts: When pure, aluminium is quite destitute of taste or odor.

Taste.

Deville: Pure aluminium has no taste, but the impure and odorous metal may have a taste like iron, in any case only very slight.

Malleability.

Deville: Aluminium may be forged or rolled with as much perfection as gold or silver. It is beaten into leaves as easily as they, and a very experienced gold beater, M. Rousseau, has made leaves as fine as those of gold or silver, which are put up in books. I know of no other useful metal able to stand this treatment. Before rolling a bar of aluminium it is well to prepare the metal by forging it on all sides, and commencing work with a hammer. Aluminium is tempered at a very low red heat; or the plate is heated just until the black trace left on its surface by a drop of oil put there and which is carbonized has entirely disappeared.

Mallet: With absolutely pure aluminium the malleability was undoubtedly improved, the metal yielding easily to the hammer, bearing distortion well, and flattening in two or three directions without cracking. It seemed to be sensibly less hardened by hammering than the ordinary metal of commerce.

'Chemical News,' 1859: M. Degousse has succeeded in beating aluminium into leaves as thin as those obtained of gold or silver. The operation is attended with a certain difficulty, and it is necessary to temper the metal frequently. This cannot be done, however, in the ordinary manner as with gold or silver; only a very slight heat must be employed. The beating is done as usual. These thin aluminium leaves can be substituted for silver leaf. They have a less brilliant color, but are much more durable, and may be employed advantageously for decorative purposes. A very thin leaf will burn like paper when made into a roll, with a brilliant white flame.

Kerl & Stohman: Aluminium may be rolled as easily as other metals, but it must be annealed oftener. The annealing of objects made of it is not more difficult than that of other metals. The moment the metal begins to glow its annealing is complete. Those metal-workers who are anxious about the exact point of time can rub the top of the article to be annealed with a lump of fat, the disappearance of the fat shows the moment in which the object is to be removed from the annealing oven. Aluminium can also be pressed or

PHYSICAL PROPERTIES OF ALUMINIUM. 59

stamped into all forms of hollow and round vessels, in a stamping press. But there must be used a kind of varnish of 4 parts of oil of turpentine and 1 part of stearic acid.

Bell Bros.: Aluminium can be beaten out, hot or cold, to the same extent and as perfectly as gold and silver, and may be rolled in much the same way. Thin leaves may be used in the same manner as gold and silver leaf. Covered iron ingot moulds serve best for casting bars of the metal to be rolled. Aluminium quickly loses its temper, and therefore requires frequent reheating at a dull red heat; when the plates are very thin this demands great attention.

Mierzinski: The extensibility of aluminium is quite high, standing near to gold and silver. It may easily be beaten out or rolled without tearing. In beating to leaf it should at first be warmed only to 100° or 150°, an actual glowing heat has proved to be very unsuitable. Such leaves are especially suitable for showing the characteristic qualities of the metal; for instance, it dissolves with extraordinary quickness in caustic alkali, leaving the iron, which is always present. This leaf is also very combustible, even in a gas flame, burning with a brilliant, sparkling light; the resulting Al^2O^3 is melted, and as hard as corundum. While water does not appear to be decomposed by aluminium in compact masses at 100°, yet it does so when in the extremely attenuated form of leaf. In pure, boiling water the leaf slowly evolves hydrogen, after several hours the leaves are half gone, being changed into hydrated alumina. Aluminium leaf was first made by C. Falk & Co., Vienna.

Ductility.

Deville: Aluminium behaves very well at the drawing plate. M. Vangeois obtained in 1855, with a metal far from being pure, wires of extreme tenuity, which were used to make aluminium passementere. However, the metal deteriorates much in the operation, and the threads become flexible again only after an annealing very delicately performed, because of the fineness of the threads and the fusibility of the metal. The heat of the air coming from the top of the chimney over an Argand burner is sufficient to anneal them.

Bell Bros.: Aluminium is easily drawn into wire. Run the metal into an open mould, so as to form a flat bar of about one-half inch section, the edges of which are beaten very regularly with a hammer. The diameter should be very gradually reduced at first, with frequent heating. When the threads are required very fine the heating becomes a very delicate operation, on account of the fineness of the threads and the fusibility of the metal.

PHYSICAL PROPERTIES OF ALUMINIUM. 61

ELASTICITY—TENACITY—HARDNESS.

Deville: The elasticity of aluminium, according to M. Wertheim, is sensibly the same as that of silver; its tenacity is also nearly the same. The moment after being cast it has the hardness of virgin silver; when it has been worked it resembles that of soft iron, becomes elastic by becoming much more rigid, and gives the sound of steel when dropped on a hard body.

Mallet: Absolutely pure aluminium was distinctly softer than before purification. Hence its fracture was not easily observed, but seemed to be very fine grained with some appearance of fibrous silkiness. It seemed to be sensibly less hardened by hammering than the ordinary metal of commerce.

Fremy: Aluminium just cast is scratched by a wire or edge of silver, but by hammering it becomes as hard as iron and elastic. The tenacity of aluminium wire is between that of zinc and tin, but by hammering it attains that of hardened copper. When cast carefully it can be easily filed without fouling the tool.

Kerl & Stohman: Aluminium resists the action of the engraving tool, which slides upon the surface of the metal as upon hard glass. But as soon as one uses the varnish of 4 parts of oil of turpentine and 1 of stearic acid, or some olive oil mixed with rum, the tool cuts into it like pure copper.

Mierzinski: The tenacity of aluminium is very remarkable, and, according to Barlow, is 1892 kilos per square centimetre; the extensibility 2.5 per cent.

W. H. Barlow :* A bar of aluminium three feet long and one-quarter inch square was obtained, and different parts of it subjected to tests for tension, compression, and transverse strain, elasticity, elastic range, and ductility. It will be seen on reference to the results that the weight of a cubic inch was 0.0275 pound, showing a specific gravity of 2.688, and its ultimate tensile strength was about twelve tons per square inch. The range of elasticity is large, the extreme to the yielding point being one-two hundredth of the length. The modulus of elasticity is 10,000. The ductility in samples two inches long was 2.5 per cent. Taking the tensile strength of the metal in relation to its weight, it shows a high mechanical value. These results are thus tabulated :—

	Weight of 1 cubic foot in pounds.	Tensile strength per sq. in. in pounds.	Length of a bar able to support its weight, in feet.
Cast Fe	444	16,500	5351
Bronze	525	36,000	9893
Wrought Fe	480	50,000	15,000
Steel	490	78,000	23,040
Al	168	26,800	23,040

It thus appears that taking the strength of aluminium in relation to its weight, it possesses a

* Rpt. Brit. A. A. S., 1882, p. 668.

PHYSICAL PROPERTIES OF ALUMINIUM. 63

mechanical value about equal to that of steel of 35 tons per square inch tensile strength.

Mierzinski: Kamarsch (Dingler 172, 55) obtains the following results as the strength of aluminium wire:—

Diameter.	Tensile Strength, Grammes.			Tenacity.
Millimetres.	1st trial.	2d trial.	Mean.	Kilos per sq. millimetre.
0.225	661	653	657	12.975
0.205	524	506	515	12.255
0.160	307	311	309	12.700
0.145	246	252	249	11.845

Sonorousness.

Deville: A very curious property, which aluminium shows the more the purer it is, is its excessive sonorousness, so that a bar of it suspended by a fine wire and struck sounds like a crystal bell. M. Lissajous, who with me observed this property, has taken advantage of it to construct tuning forks of aluminium, which vibrate very well. I also tried to cast a bell, which has been sent to the Royal Institution at London at the request of my friend Rev. J. Barlow, vice-president and secretary of the institution. This bell, cast on a model not well adapted to the qualities of the metal, gives a sharp sound of considerable intensity, but which is not prolonged, as if the clapper or support hindered the sound, which, thus hindered, becomes far from agreeable. The sound produced by the ingots is, on the contrary, very pure and prolonged.

In the experiments made in Mr. Faraday's laboratory, this celebrated physicist has remarked that the sound produced by an ingot of aluminium is not simple. One can distinguish, by turning the vibrating ingot, two sounds very near together and succeeding each other rapidly, according as one or the other face of the ingot faces the observer.

Watts: Aluminium is highly sonorous, but a bell cast of it gave a sound like a cracked pot.

Density.

Deville: The density of aluminium is 2.56; by rolling this is considerably increased, so as to become 2.67, indicating a considerable approaching of the molecules to each other; which may explain the differences existing in its properties after being annealed or worked. Heated to $100°$ and cooled, it changes very little, for its specific gravity is still 2.65. The following table compares it with the other metals:—

	Sp. Gr.	Sp. Gr. Al. = 1.
Pt	21.5	8.6
Au	19.3	7.7
Pb	11.4	4.8
Hg	10.5	4.2
Cu	8.9	3.6
Fe	7.8	2.9
Sn	7.3	2.8
Zn	7.1	2.8
Al	2.5	1.0

Since the metal has been in commerce it has been sold at a high price; at present (1859) it can be bought in large quantities at 300 fr. per kilo; it is, therefore, much dearer than silver. But, because of the difference in their densities, for equal volumes of aluminium and silver, the value of the former must be divided by 4 in order to compare them; making a volume of aluminium much cheaper than an equal volume of silver, while, besides, it is much stronger. So, to-day, Al may be considered as costing 75 fr. to Ag 220 fr.

Mallet: The specific gravity of absolutely pure aluminium was carefully determined at 4° C., and the mean of three closely agreeing observations gave 2.583.

Fusibility.

Deville: Aluminium melts at a temperature higher than that of zinc, lower than that of silver, but approaching nearer to that of zinc than silver. It is, therefore, quite a fusible metal.

Mallet: It seems that pure aluminium is a little less fusible than the commercial metal.

Mierzinski: The melting point of aluminium can be taken as about 700° C.

Fixity.

Deville: Aluminium is absolutely fixed, and loses no part of its weight when it is violently heated in a forge fire in a carbon crucible.

Watts: Aluminium heated in a closed vessel does not exhibit the slightest tendency to volatilize.

Fremy: Aluminium is fixed at all temperatures.

Electric Conductivity.

Deville: Aluminium conducts electricity with great facility, so that it may be considered as one of the best conductors known, and perhaps equal to silver. I found by Wheatstone's Bridge that it conducts eight times better than iron. M. Buff has arrived at results evidently different from mine because we have not taken the same ground of comparison. The difference is due, without doubt, to the metal which he employed containing, as is easily found in many specimens, a little cryolite and fusible materials, the density of which is near that of the metal, and which were employed in producing it. The complete separation of the metal and flux is a difficult mechanical operation, but which is altogether avoided by using a volatile flux. This is a condition which must be submitted to in order to get the metal absolutely pure.

'Jahresb. der Chemie,' 1881, p. 94: Aluminium

thus compares with copper and magnesium in electric conductivity:—

	At 0°.	At 100°.
Cu	45.74	33.82
Mg	24.47	17.50
Al	22.46	17.31

After Al come red brass, Cd, yellow brass, Fe, Zn, Pb, Ag, Sb, Bi, in the order given.

Fremy: The electric conductivity of aluminium is 51.5, copper being 100; or 33.74, silver being 100.

THERMAL CONDUCTIVITY.

Deville: It is generally admitted that conductivity for heat and electricity correspond exactly in the different metals. A very simple experiment made by Mr. Faraday in his laboratory seems to place aluminium very high among metallic conductors. He found that it conducted heat better than silver or copper.

Watts: Aluminium conducts heat better than silver.

'Jahresb. der Chemie,' 1881, p. 94: Aluminium has the following conductivity for heat:—

	At 0°.	At 100°.
Cu	0.7198	0.7226
Mg	0.3760	0.3760
Al	0.3435	0.3619

After Al come red brass, Cd, yellow brass, Fe, Zn, Pb, Ag, Sb, Bi in the order given.

Mierzinski: No less remarkable than the conductivity of aluminium for electricity is that for heat. According to Calvert and Johnson (Dingler, 153, 285), that of silver being 1000, aluminium is 665.

Specific Heat.

Deville: According to the experiments of M. Regnault, the specific heat of aluminium corresponds to its equivalent 13.75, from which we may conclude that it must be very large when compared with all the other useful metals. One can easily perceive this curious property by the considerable time which it takes an ingot of the metal to get cold. We might even suggest that a plate of aluminium would make a good chafing dish. Another experiment makes this conclusion very evident. M. Paul Morin had the idea to use aluminium for a plate on which to cook eggs, the sulphur of which attacked silver so easily; and he obtained excellent results. He noticed, also, that the plate kept its heat a much longer time than the silver one. This exceptional property should be utilized for something.

Mallet: The specific heat of absolutely pure aluminium was 0.2253, therefore the atomic heat is 0.2253 times 27.02 or 6.09.

Fremy: The specific heat of aluminium is

0.2181; larger than that of any other useful metal, which accords with its small atomic weight.

MAGNETISM.

Deville: I have found, as also MM. Poggendorff and Reiss, that aluminium is very feebly magnetic.

CRYSTALLINE FORM.

Deville: Aluminium often presents a crystalline appearance when it has been cooled slowly. When it is not pure the little crystals which form are needles, and cross each other in all directions. When it is almost pure it still crystallizes by fusion, but with difficulty, and one may observe on the surface of the ingots hexagons which appear regularly parallel along lines which centre in the middle of the polygon. It is an error to conclude from this observation that the metal crystallizes in the rhombohedral system. It is evident that a crystal of the regular system may present a hexagonal section; while, on the other hand, in preparing aluminium by the battery at a low temperature, I have observed complete octahedrons, which were impossible of measurement, it is true, but their angles appeared equal.

PART IV.

CHEMICAL PROPERTIES OF ALUMINIUM.

REMARK: Unless specifically stated otherwise, the properties here mentioned are those of the pure metal and not of the commercial, the impurities of which generally modify the properties of the aluminium more or less.

ACTION OF AIR.

Deville: Air, wet or dry, has absolutely no action on aluminium. No observation which has come to my knowledge is contrary to this assertion, which may easily be proved by any one. I have known of beams of balances, weights, plaques, polished leaf, reflectors, etc., of the metal exposed for months to moist air and sulphur vapors, and showing no trace of alteration. We know that aluminium may be melted in the air with impunity. Therefore air and also oxygen cannot sensibly affect it; it resisted oxidation in the air at the highest heat I could produce in a cupel furnace, a heat much higher than that required for the assay of gold. This experiment is interesting, especially when the metallic button is covered with a layer of oxide which tarnishes it, the expansion of the metal causes small branches to shoot from its surface, which are very brilliant, and do not lose their lustre in spite of the oxidizing atmosphere. M. Wöhler has also observed this property on trying to melt the metal with a blowpipe. M. Peligot has profited by it to cupel aluminium. I have seen buttons of impure metal cupelled with lead and become very malleable.

With pure aluminium the resistance of the metal to direct oxidation is so considerable that at the melting point of platinum it is hardly appreciably touched, and does not lose its lustre. It is well known that the more oxidizable metals take this property away from it. But silicon itself, which is much less oxidizable, when alloyed with it makes it burn with great brilliancy, because there is formed a silicate of aluminium.

Watts: Aluminium may be heated intensely in a current of air in a muffle without undergoing more than superficial oxidation. When heated as foil with a splinter of wood in a current of oxygen it burns with a brilliant, bluish-white light.

'Chemical News,' 1859: Wöhler finds that aluminium leaf burns brightly in air and in oxygen with a brilliant light. The Al^2O^3 formed is as hard as corundum. Wire burns in oxygen like iron wire, but the combustion cannot continue because the wire fuses.

Mierzinski: Aluminium does not change at a somewhat high temperature in the air; but if heated to whiteness it burns, with the production of strong light, to Al^2O^3, which covers the surface of the bath.

Action of Water (H^2O).

Deville: Water has no action on aluminium, either at ordinary temperatures, or at 100°, or at a red heat bordering on the fusing point of the metal. I boiled a fine wire in water for half an hour and it lost not a particle in weight. The same wire was put in a glass-tube heated to redness by an alcohol lamp and traversed by a current of steam, but after several hours it had not lost its polish, and had the same weight. To obtain any sensible action it is necessary to operate at the highest heat of a reverberatory furnace, a white heat. Even then the oxidation is so feeble that it develops only in spots, producing almost inappreciable quantities of Al^2O^3. This slight alteration and the analogies of the metal allow us to admit that it decomposes water, but very feebly. If, however, metal produced by M. Rose's method was used, which is almost unavoidably contaminated with slag composed of chlorides of aluminium and sodium, the Al^2Cl^6, in presence of water, plays the part of an acid towards aluminium, disengaging hydrogen with the formation of a sub-

CHEMICAL PROPERTIES OF ALUMINIUM.

chlorhydrate of alumina, whose composition is not known, and which is soluble in water. When the metal thus tarnishes in water one may be sure to find chlorine in the water on testing it with nitrate of silver.

Mierzinski: Cold and warm water have no influence on aluminium even if it is heated to redness.

'Chemical News,' 1859: Aluminium leaf will slowly decompose water at 100°; at first it takes a bronze color, and after boiling some hours it becomes translucent.

ACTION OF HYDROGEN SULPHIDE AND SULPHUR (H^2S and S).

Deville: Sulphuretted hydrogen exercises no action on aluminium, as may be proved by leaving the metal in an aqueous solution of the gas. In these circumstances almost all the metals, and especially silver, blacken with great rapidity. Sulph-hydrate of ammonia may be evaporated on an aluminium leaf, leaving on the metal only a deposit of sulphur which the least heat drives away.

Aluminium may be heated in a glass tube to a red heat in vapor of sulphur without altering the metal. This resistance is such that in melting together polysulphide of potassium and some aluminium containing copper or iron, the latter are attacked without the aluminium being sensibly affected. Unhappily, this method of purification may not be employed because of the protection which aluminium exercises over foreign metals. Under the same circumstances gold and silver dissolve up very rapidly. However, at a high temperature I have observed that it combines directly with sulphur to give Al^2S^3. These properties varying so much with the temperature form one of the special characteristics of the metal and its alloys.

Fremy: H^2S is without action on aluminium, acting towards it as towards the sulphides of iron, zinc, or copper. It is true that aluminium decomposes Ag^2S, but it sets the sulphur at liberty and combines with the silver. These facts are in accordance with the resistance the metal offers to free sulphur.

Sulphuric Acid (H^2SO^4).

Deville: Sulphuric acid, diluted in the proportion most suitable for attacking the metals which decompose water, has no action on aluminium; and contact with a foreign metal does not help, as with zinc, the solution of the metal, according to M. de la Rive. This singular fact tends to remove aluminium considerably from those metals. To establish it better, I left for several months some globules weighing only a few milligrammes in contact with weak H^2SO^4, and they showed no visible alteration; however the acid gave a faint precipitate with aqua ammonia.

Fremy: H^2SO^4, dilute or concentrated, exercises in the cold only a very slight sensible action on aluminium, the pure metal is attacked more slowly than when it contains foreign metals. The presence of silicon gives rise to a disengagement of SiH^4, which communicates to the hydrogen set free a tainted odor. Concentrated H^2SO^4 dissolves it rapidly with the aid of heat, disengaging sulphurous acid gas (SO^2).

Nitric Acid (HNO^3).

Deville: Nitric acid, weak or concentrated, does not act on aluminium at the ordinary temperature. In boiling HNO^3 the solution takes place, but with such slowness that I had to give up this mode of dissolving the metal in my analyses. By cooling the solution all action ceases. M. Hulot has obtained good results on substituting aluminium for platinum in the Grove battery.

Hydrochloric Acid (HCl).

Deville: The true solvent of aluminium is HCl, weak or concentrated; but, when the metal is perfectly pure, the reaction takes place so slowly that M. Favre, of Marseilles, had to give up this way of attack in determining the heat of a combination of the metal. But impure aluminium is dissolved very rapidly. At a very low temperature gaseous HCl attacks the metal and changes it into Al^2Cl^6. Under these circumstances iron does not seem to alter; able, no doubt, to resist by covering itself with a very thin protecting layer of $FeCl^2$. This experiment would lead me to admit that it is the HCl and not the water which is decomposed by aluminium; and, in fact, the metal is attacked more easily as the acid is more concentrated. This explains the difference of the action of solutions of HCl and H^2SO^4, the latter being almost inactive. This reasoning applies also to tin.

When the metal contains silicon, it disengages hydrogen of a more disagreeable smell than that given out by iron under similar circumstances. The reason of this is the production of that remarkable body recently discovered by MM. Wöhler and Buff—SiH^4. When the proportion of silicon is small, the whole is evolved as gas; when increased a little, some remains in solution with the aluminium, and then it requires great care to separate the metal exactly, even when the solution is evaporated to dryness. If 3 to 5 per cent. of Si is present, it remains insoluble mixed with a little SiO^2, as has been cleverly proven by Wöhler and Buff by the action of hydrofluoric acid, which dissolves the SiO^2 with evolution of H without attacking the Si itself. On dissolving commercial aluminium there is sometimes obtained a black, crystalline residue, which, separated on a filter and dried at 200° to 300° takes fire in places; this residue is Si mixed with some SiO^2. The presence of Si augments very much the facility with which Al is attacked by HCl.

Mierzinski: If HCl is present in a mixture of acids, it begins the destruction of the metal. HI, HBr, and HF act similarly to HCl.

POTASH, SODA, AND LIME (KOH, NaOH, $Ca(OH)^2$).

Deville: Alkaline solutions act with great energy on the metal, transforming it into aluminate of potash or soda, setting free hydrogen. However, it is not attacked by KOH or NaOH in fusion; one may, in fact, drop a globule of the pure metal into melted caustic soda raised almost to a red heat in a silver vessel, without observing the least disengagement of hydrogen. Silicon, on the contrary, dissolves with great energy under the same circumstances. I have employed melted NaOH to clean siliceous aluminium. The piece is dipped into melted NaOH kept almost at red heat. At the moment of immersion several bubbles of H disengage from the metallic surface, and when they have disappeared, all the Si of the superficial layer of Al has been dissolved. It only remains to wash well with water and dip it into nitric acid, when the aluminium takes a beautiful mat.

Mallet: The pure metal presents greater resistance to the prolonged action of alkalies than the impure.

Mierzinski: Lime-water acts similarly to NaOH or KOH, with the difference that the resulting calcium compound is precipitated.

Aqua Ammonia (NH^4OH).

Deville: Aqua ammonia acts only feebly on aluminium, producing a little Al^2O^3, which, according to a very curious observation of Wöhler, has the property of partly dissolving in the ammonia. In an atmosphere in which ammonia was present, the metal did not lose its lustre, which is easily explained, because it is only in contact with water that the oxidization of the metal takes place, with disengagement of hydrogen.

Organic Acids, Vinegar, etc.

Deville: Weak acetic acid acts on aluminium in the same way as H^2SO^4, *i. e.*, in an inappreciable degree or with extreme slowness. I used for the experiment acid diluted to the strength of strongest vinegar. M. Paul Morin left a plaque of the metal a long time in wine which contained tartaric acid in excess and acetic acid, and found the action on it quite inappreciable. The action of a mixture of acetic acid and NaCl in solution in pure water on pure aluminium is very different, for the acetic acid replaces a portion of the HCl existing in the NaCl, rendering it free. However, this action is very slow on the Al, especially if it is pure.

The practical results flowing from these observations deserve to be clearly defined, because of the applications which may be made of aluminium to culinary vessels. I have observed that the tin so often used, and which each day is put in contact with NaCl and vinegar, is attacked much more rapidly than aluminium under the same circumstances. Although the salts of tin are very poisonous, and their action on the economy far from being negligible, the presence of tin in our food passes unperceived because of its minute quantity. Under the same circumstances, aluminium dissolves in less quantity; the acetate of Al formed resolves itself on boiling into insoluble Al^2O^3 or an insoluble sub-acetate, having no more taste or action on the body than clay itself. It is for that reason, and because it is known that the salts of the metal have no appreciable action on the body, that aluminium may be considered as an absolutely harmless metal.

Solutions of Metallic Salts.

Deville: The action of any salt whatever may be easily deduced from the action of the acids on the metal. We may, therefore, predict that in acid solutions of sulphates and nitrates aluminium will precipitate no metal, not even silver, as Wöhler has observed. But the hydrochloric solutions of the same metals will be precipitated, as MM. Tissier have shown. Likewise, in alkaline solutions, Ag, Pb, and metals high in the classification of the elements are precipitated.

It may be concluded from this that to deposit aluminium on other metals by means of the battery, it is always necessary to use acid solutions in which HCl, free or combined, should be absent. For similar reasons the alkaline solutions of the same metals cannot be employed, although they give such good results in plating common metals with gold and silver. It is because of these curious properties that gilding and silvering aluminium are so difficult. M. Paul Morin and I have often tried a bath of basic sulphide of gold, or hyposulphite of silver, with a large excess of sulphurous acid, with no good results. But M. Mourey, who has already rendered great services in galvanoplasty, readily gilds and silvers aluminium for commerce with astonishing skill when we consider the short time he has had to study this question. I know also that M. Christofle has gilded it, but I am entirely ignorant of the processes employed by these gentlemen. The coppering of aluminium by the battery is effected very easily by means of M. Hulot's process. He uses simply a bath of acid sulphate of copper. The layer of copper, if well prepared, is very solid.

All that I have said on the subject of the action

CHEMICAL PROPERTIES OF ALUMINIUM. 81

of metallic salts is true only for pure aluminium. Impure metal, especially if it contains iron or sodium, acts then in producing in the copper salts with which I operate a deposit of metallic Cu. But this phenomenon, even in the most unfavorable cases, is produced very slowly, and if a leaf of aluminium is used one may see at the end of several weeks the texture of the metal etched with red fibres, as if the Fe and Al were only in juxtaposition, and the ferruginous fibres acted alone. Moreover, the deposit is only local, and little by little becomes complete; but it is slower as the metal is purer.

Mierzinski: Silver is precipitated by Al from a nitrate solution, feebly acid or neutral, in dendrites; the separation begins after six hours; from an ammoniacal solution of $AgCl$ or Ag^2CrO^4, Al precipitates the metal immediately as a crystalline powder.

From $CuSO^4$ or $Cu(NO^3)^2$ solution, Al separates Cu only after two days, in dendrites or octahedra; from the latter it also precipitates a basic salt as a green, insoluble powder; from a $CuCl^2$ solution the Cu falls immediately; somewhat slower from solution of acetate of copper. The sulphate or nitrate solution behaves similarly if a little KCl is added to it, and the precipitation is complete in presence of excess of Al.

From Hg^2Cl^2, Hg^2Cy^2, and $Hg^2(NO^3)^2$, Hg separates first, and then forms an amalgam with the Al which decomposes water at ordinary temperatures or oxidizes in the air with development of much heat; the same qualities are possessed by the amalgam formed by warming the two metals together in an atmosphere of CO^2.

From $Pb(NO^3)^2$ and $Pb(C^2H^3O^2)^2$ the metal separates slowly in crystals; from $PbCl^2$ immediately; an alkaline solution of $PbCrO^4$ gives Pb and Cr^2O^3.

From an alcoholic solution of $HgCl^2$ the Hg is precipitated much quicker with a gentle heat. Al also reduces Hg from a solution of HgI^2 in KI. Al separates Hg from $HgCl^2$ vapors, and Al^2Cl^6 deposits in the cooler part of the tube, the remaining Al being melted by the heat of the reaction. Al acts likewise toward melted AgCl, the silver set free being melted by the heat of the reaction. Zn is easily thrown down from alkaline solution.

Fremy: Aluminium decomposes a very large number of metallic solutions, which takes place especially easily if the solution is made alkaline or ammoniacal. Acid, and especially neutral solutions, are less favorable for the experiment. All the chlorides, excepting KCl and NaCl, are reduced by it. Al^2Cl^6 is no exception, for the solution is decomposed with evolution of hydrogen. Aluminium easily resists solutions of NaCl and alum separately, but dissolves in a mixed solution of these two salts. In alkaline solution, the metals are precipitated because of the facility with which aluminates of the alkalies are formed.

Watts (2d Supplement): The action of aluminium on metallic solutions is as follows: Cu is precipitated from copper salts; Pb is slowly precipitated from lead salts; Ag is precipitated from a slightly acid or neutral solution of $AgNO^3$; Zn is readily precipitated from zinc salts.

NITRE.

Deville: Aluminium may be melted in nitre without undergoing the least alteration, the two materials rest in contact without reacting, even at, a red heat, at which temperature the salt is plainly decomposed, disengaging oxygen actively. But if the heat is pushed to the point where nitrogen itself is disengaged, there the nitre becomes potassa, a new affinity becomes manifest, and the phenomena change. The metal then combines rapidly with the K^2O to give aluminate of potash. The accompanying phenomenon of flagration often indicates a very energetic reaction. Aluminium is continually melted with nitre at a red heat to purify it by the oxygen disengaged, without any fear of loss. But it is necessary to be very careful in doing it in an earthen crucible. The SiO^2 of the crucible is dissolved by the nitre, the glass thus formed is decomposed by the aluminium, and the silicide of aluminium formed is then very oxidizable, especially in the presence of alkalies. The purification by nitre ought to be made in an iron crucible well oxidized by nitre inside.

Fremy: At the melting point, aluminium is not

attacked by nitre; this property has been at times utilized to oxidize and then remove the metals alloyed with it, but it is now demonstrated that this mode of purification is very imperfect.

Mierzinski: Heated to redness with nitre, aluminium burns with a fine blue flame.

SILICATES AND BORATES.

Deville: By treating silicates and borates with aluminium silicon and boron may be obtained. The process is described at the end of Deville's book, but is too long and foreign to the subject in hand to be given here.

Tissier: Aluminium melted in an ordinary white glass vessel oxidizes itself at the expense of the SiO^2, setting free silicon, and the alumina formed combines with the alkali forming an aluminate. In experiments which we have made, the metal became covered with a thin layer of silicon, while the metal which remained underneath was still malleable and did not appear to be combined with Si.

FLUORSPAR.

Tissier: This salt is without action on the metal and makes its best flux, especially so because of the property which it has of dissolving the alumina with which the metal may be contaminated and which encrusts little globules. The fluorspar, by dissolving this crust, facilitates their reunion.

Phosphate of Lime.

Tissier: We have heated to white heat a mixture of pure $Ca^3(PO^4)^2$ and aluminium leaf, without the metal losing its metallic appearance. This material thus appears to have no action on the metal.

Sodium Chloride (NaCl) and Chlorides.

Deville: A solution of sodium or potassium chloride, in which is put a pure aluminium wire, seemed to me to exercise no sensible action on the metal, either cold or warm. It is not the same with the other metallic chlorides, and we may state that, as a general rule, these are decomposed by aluminium with greater facility as the metal which they contain belongs to a higher order. The chlorhydrate of Al itself dissolves aluminium forming a sub-chlorhydrate with evolution of hydrogen.

Tissier: NaCl is employed as a flux for Al in remelting it. It does not possess the property, like CaF^2, of dissolving the Al^2O^3, and has the inconvenience of producing with the clay of the crucible a sensible quantity of Al^2Cl^6, which may on contact with the air act in promoting the loss of a certain quantity of metal.

Metallic Oxides.

Tissier: We made our experiments in this way: The Al leaf was mixed carefully with the oxide on which we experimented, then the mixture was placed in a small porcelain capsule and heated in a small earthen crucible which served as a muffle. Our results were as follows:—

MnO^2—Aluminium has no action on manganese dioxide.

Fe^2O^3—By heating to white heat 1 equivalent of Fe^2O^3 and 3 of Al, the reaction took place with detonation, and by heating sufficiently we obtained a metallic button, well melted, and containing 69.3 per cent. Fe and 30.7 per cent. Al, being as hard and brittle as cast-iron. Its composition is nearly AlFe. It would thus appear that the decomposition of Fe^2O^3 will not pass the limit where the quantity of iron reduced is sufficient to form with the aluminium the alloy AlFe.

ZnO: A mixture of aluminium leaf and zinc oxide heated to whiteness did not appear to present the least indication of decomposition.

PbO: We mixed 2 equivalents of litharge with 1 of aluminium, and heated the mixture slowly up to white heat, when the Al reacted on the PbO with such intensity as to produce a strong detonation. We made an experiment with 50 grammes of PbO and 2.9 grammes of Al leaf, when the crucible was broken to pieces and the doors of the furnace blown off.

CuO: 3 grammes of black oxide of copper mixed with 1.03 grammes of aluminium detonated producing a strong explosion when the heat reached whiteness.

Mierzinski: Aluminium reduces CuO and PbO with explosion, Fe^2O^3 only in part, forming the alloy AlFe. ZnO and MnO are not reduced by aluminium.

Bekétoff:* He reduced baryta (BaO) with metallic aluminium in excess, and obtained alloys of aluminium and barium containing in one case 24 per cent. in another 33 per cent. of Ba.

Animal Matters.

Deville: Among the animal matters produced by the organism, some are acid, as sweat. These appear to have no sensible action on aluminium. Alkaline materials, as the saliva, have a greater tendency to oxidize it, but the whole effect produced is insignificant. M. Charrière has made for a patient on whom he practised tracheotomy a small tube of the metal, which remained almost unaltered although in contact with purulent matter. After a long time a little alumina was formed on it, hardly enough to be visible.

* Bull. de la Soc. Chem., 1857, p. 22.

Miscellaneous Agents.

Tissier: Only 2.65 grammes of aluminium introduced into melted red hot sodium sulphate (Na^2SO^4) decomposed that salt with such intensity that the crucible was broken into a thousand pieces, and the door of the furnace blown to a distance. Heated to redness with alkaline carbonate, the Al was slowly oxidized at the expense of the CO^2, C was set free, and an aluminate formed. The reaction takes place without deflagration.

Mierzinski: Heated to redness with potassium or sodium sulphate, aluminium gives a strong detonation. Potassium carbonate quickly destroys the metal with separation of carbon. Hydrogen, nitrogen, sulphur, and carbon are without any influence on aluminium, but chlorine, iodine, bromine, and fluorine attack it rapidly.

General Observations on the Properties of Aluminium.

Deville: Aluminium, at a low temperature, conducts itself as a metal which can give a very weak base; in consequence, its resistance to acids, HCl excepted, is very great. It conducts itself with the alkalies as a metal capable of giving a quite energetic acid, it being attacked by K^2O and Na^2O dissolved in water. But, this affinity is still insufficient to determine the decomposition of melted KOH. For a stronger reason it does not decompose metallic oxides at a red heat. This is why, in the muffle, the alloy of aluminium and copper gives black CuO, and this also accounts for the alloy of aluminium and lead being capable of being cupelled. But, by a strange exception, and which does not appertain solely, I believe, to aluminium, as soon as the heat is above redness the affinities are quickly inverted, and the metal takes all the properties of silicon, decomposing the oxides of lead and copper with the production of aluminates.

From all the experiments which have been reported and from all the observations which have been made, we can conclude that aluminium is a metal which has complete analogies with no one of the simple bodies which we consider metals. In 1855 I proposed to place it along side of chromium and iron, leaving zinc out of the group with which aluminium had been until then classed. Zinc is placed very well beside magnesium, there being intimate analogies between these two volatile metals. There may be found at the end of a memoir which M. Wöhler and I published in the 'Compt Rendue' and the 'Ann. de Chem. et de Phys.' the reasons why we are tempted to place aluminium near to silicon and boron in the carbon series, on grounds analogous to those on which antimony and arsenic are placed in the nitrogen series.

PART V.

METALLURGY OF ALUMINIUM.

As has been remarked in the historical section, Davy was the first to try to isolate aluminium. His attempts were unsuccessful. The next chemist to publish an account of attempts in this direction was Oerstedt, who published a paper in 1824 in a Swedish periodical.* Oerstedt's original paper is thus translated into Berzelius' 'Jahresbericht:'†

"Oerstedt mixes calcined and pure alumina, quite freshly prepared, with powdered charcoal, puts it in a porcelain retort, ignites and leads Cl gas through. The coal then reduces the alumina, and there results Al^2Cl^6 and CO, and perhaps also some phosgene, $COCl^2$; the Al^2Cl^6 is caught in the condenser and the gases escape. The Al^2Cl^6 is white, crystalline, melts about the temperature of boiling water, easily attracts moisture, and evolves heat when in contact with water. If it is mixed with a concentrated potassium amalgam and heated quickly, it is transformed; there results KCl, and

* Oversigt over det K. Danske Videnskabemes Selkabs Forhandlingar og dets Medlemmers Arbeider. May, 1824, to May, 1825, p. 15.

† Berz. Jahresb. der Chemie, 1827, vi. 118.

the aluminium unites with the mercury. The new amalgam oxidizes in the air very quickly, and gives as residue when distilled in a vacuum a lump of metal resembling tin in color and lustre. In addition, Oerstedt found many remarkable properties of the metal and of the amalgam, but he holds them for a future communication after further investigation."

I have not been able to find any other paper by Oerstedt, but the next advance in the science is by Wöhler, and all agree in naming him as the true discoverer of the metal. The following is taken from Poggendorf.*

Wöhler reviews the article which we have just given, and then continues as follows:—

"I have repeated this experiment of Oerstedt, but achieved no very satisfactory result. By heating potassium amalgam with Al^2Cl^6 and distilling the product, there remained behind a gray melted mass of metal, but which, by raising the heat to redness, went off as green vapor and distilled as pure potassium. I have therefore looked around for another method or way of conducting the operation, but, unpleasant as it is to say it, the reduction of the aluminium fails each time. Since, however, Herr Oerstedt remarks at the end of his paper that he did not regard his investigations in aluminium as yet ended, and already several years

* Pogg. Ann., 1827, ii. 147.

have passed since then, it looks as if I had taken up one of those researches begun auspiciously by another (but not finished by him) because it promised new and splendid results. I must remark, however, that Herr Oerstedt has indirectly by his silence encouraged me to try to attain to further results myself. Before I give the art how one can quite easily reduce the metal, I will say a few words about Al^2Cl^6 and its production.

" I based the method of reducing aluminium on the reaction of Al^2Cl^6 on potassium, and on the property of the metal not to oxidize in water. I warmed in a glass retort a small piece of Al^2Cl^6 with some potassium, and the retort was shattered with a strong explosion. I tried then to do it in a small platinum crucible, in which it succeeded very well. The reaction is always so violent that the cover must be weighted down, or it will be blown off; and at the moment of reduction, although the crucible be only feebly heated from outside, it suddenly glows inside, and the platinum is almost torn by the sudden shocks. In order to avoid any mixture of platinum with the reduced aluminium, I next made the reduction in a porcelain crucible and succeeded then in the following manner: Put in the bottom of the crucible a piece of potassium free from carbon and oil, and cover this with an equal volume of pieces of Al^2Cl^6. Cover, and heat over a spirit lamp, at first gently, that the crucible be not broken by the production of heat inside, and then heat stronger, at last to redness. Cool; and when fully cold put it into a glass of cold water. A gray powder separates out which on nearer observation, especially in sunlight, is seen to consist of little flakes of metal. After it has separated, pour off the solution, filter, wash with cold water, and dry; this is the aluminium."

In reality this powder possessed no metallic properties, and, moreover, it contained potassium and Al^2Cl^6, which gave to it the property of decomposing water at 100°. To avoid the loss of Al^2Cl^6 by volatilization at the high heat developed during the reaction, Liebig afterwards made the vapor of Al^2Cl^6 pass slowly over some potassium placed in a long glass tube. This device of Liebig is nearly the arrangement which Wöhler adopted later, in 1845, and which gave him much better results. The following is Wöhler's second paper, published in 1845:—*

" On account of the violent incandescence with which the reduction of Al^2Cl^6 by potassium is accompanied, this operation requires great precautions, and can be carried out only on a small scale. I took for the operation a platinum tube, in which I placed Al^2Cl^6 and near it some potassium in a platinum boat. I heated the tube gently at first, then to redness. But the reduction may also be done by putting potassium in a small crucible

---

* Liebig's Annalen, 53, 422.

which is placed inside a larger one, and the space between the two filled with Al^2Cl^6. A close cover is put over the whole and it is heated. Equal volumes of potassium and Al^2Cl^6 are the best proportions to employ. After cooling, the tube or crucible is put in a vessel of water. The metal is obtained as a gray, metallic powder, but on closer observation one can see even with the naked eye small tin-white globules some as large as pins' heads. Under a microscope magnifying two hundred diameters the whole powder resolves itself into small globules, several of which may sometimes be seen sticking together, showing that the metal was melted at the moment of reduction. A beaten-out globule may be again melted to a sphere in a bead of borax or salt of phosphorus, but rapidly oxidizes during the operation, and if the heat is continued, disappears entirely, seeming either to reduce boron in the borax bead or phosphorus or P^2O^5 in the salt of phosphorus bead. I did not succeed in melting together the pulverulent aluminium in a crucible with borax, at a temperature which would have melted cast iron, for the metal disappeared entirely and the borax became a black slag. It seems probable that aluminium, being lighter than molten borax, swims on it and burns. The white metallic globules had the color and lustre of tin. It is perfectly malleable and can be hammered out to the thinnest leaves. Its specific gravity, determined with two globules weighing 32 milligrammes, was

2.50, and with three hammered-out globules weighing 34 milligrammes, 2.67. On account of their lightness these figures can only be approximate. It is not magnetic, remains white in the air, decomposes water at 100°, not at usual temperatures, and dissolves completely in caustic potash (KOH). When heated in oxygen almost to melting, it is only superficially oxidized, but it burns like zinc in a blast-lamp flame."

These results of Wöhler's, especially the determination of sp. gr., were singularly accurate when we consider that he established them working with microscopic bits of the metal. It was just such work that established Wöhler's fame as an investigator. However, we notice that his metal differed from aluminium as we know it in several important respects, in speaking of which Deville says: "All this time the metal obtained by Wöhler was far from being pure; it was very difficultly fusible, owing without doubt to the fact that it contained platinum taken from the vessel in which it had been prepared. It is well known that these two metals combine very easily at a gentle heat. Moreover, it decomposed water at 100°, which must be attributed either to the presence of some potassium or to Al^2Cl^6, with which the metal might have been impregnated; for aluminium in presence of Al^2Cl^6 in effect decomposes water with evolution of hydrogen.

After Wöhler's paper in 1845, the next improvement is that introduced by Deville, in 1854–55, and this is really the date at which aluminium, the metal, became known and its true properties established. He first read to the Academy an account of his laboratory process, by which he obtained a pencil of the metal. The following is his account:*

"The following is the best method for obtaining aluminium chemically pure in the laboratory. Take a large glass tube about four centimetres in diameter, and put into it pure Al^2Cl^6 free from iron, and isolate it between two stoppers of amianthus (fine, silky asbestos). Hydrogen, well dried and free from air, is brought in at one end of the tube. The Al^2Cl^6 is heated in this current of gas by some lumps of charcoal, in order to drive off hydrochloric acid or sulphides of chlorine or of silicon, with which it is always impregnated. Then there are introduced into the tube porcelain boats, as large as possible, each containing several grammes of sodium, which was previously rubbed quite dry between leaves of filter paper. The tube being full of hydrogen the sodium is melted, the Al^2Cl^6 is heated and distils, and decomposes in contact with the sodium with incandescence, the intensity of which can be moderated at pleasure. The operation is ended when all the sodium has disappeared, and when the sodium chloride formed has absorbed so much Al^2Cl^6 as to be saturated with it. The Al

* Ann. de. Phys. et de Chem., xliii. 24.

METALLURGY OF ALUMINIUM.

which has been formed is held in the double chloride of sodium and aluminium, $Al^2Cl^6.2NaCl$, a compound very fusible and very volatile. The boats are then taken from the glass tube, and their entire contents put in boats made of retort carbon, which have been previously heated in dry chlorine in order to remove all silicious and ferruginous matter. These are then introduced into a large porcelain tube, furnished with a prolongation and traversed by a current of hydrogen, dry and free from air. This tube being then heated to redness, the $Al^2Cl^6.2NaCl$ distils without decomposition and condenses in the prolongation. There is found in the boats, after the operation, all the Al which had been reduced, collected in at most one or two small buttons. The boats when taken from the tube should be nearly free from $Al^2Cl^6.2NaCl$ and also from NaCl. The buttons of aluminium are united in a small earthen crucible which is heated as gently as possible, just sufficient to melt the metal. The latter is pressed together and skimmed clean by a small rod or tube of clay. The metal thus collected may be very suitably cast in an ingot mould."

The later precautions added to the above given process were principally directed towards avoiding the attacking of the crucible, which always takes place when the metal is melted with a flux, and the aluminium thereby made more or less siliceous. The next improvement was the introduction by

ALUMINIUM.

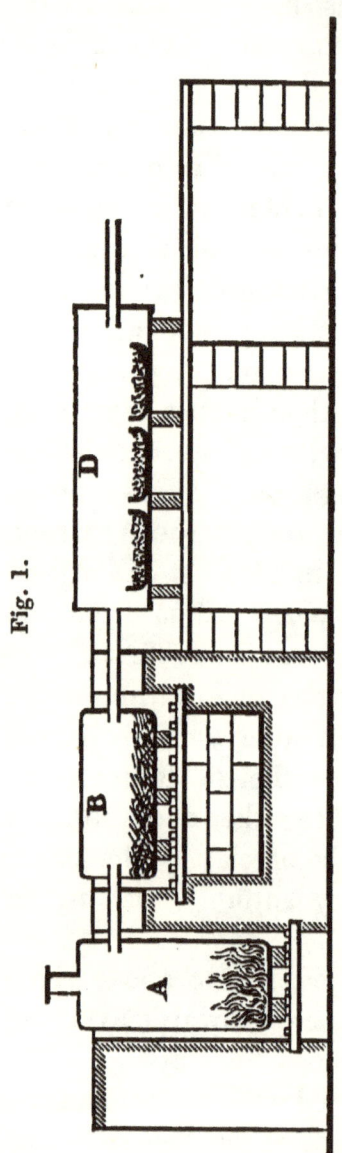

Fig. 1.

Deville of an application on a large scale of the laboratory method just described. He first put it up at the chemical works of M. du Sussex, at Javel, and later at the works of MM. Rousseau Bros., at Glacière. It has at present only an historic interest, as it was soon modified in its details so as to be almost entirely changed, but I give it here so as to show the different phases through which the industry has passed. The text is not given in full as Deville describes it, which would be unnecessary; but the condensed account gives a clear idea of the process. The full description may be found in Deville's book, or in the 'Ann. de Chem. et de Phys.' [3] xlvi. 445, where it first appeared.

The crude Al^2Cl^6, placed in the cylinder A, is vaporized by the fire and passes through the tube to the cylinder B containing 60 to 80 kilos of iron nails heated to a dull-red heat. The iron retains as relatively fixed ferrous chloride, the ferric chloride and hydrochloric acid which contaminate the Al^2Cl^6, and likewise transforms any sulphur dichloride (SCl^2) in it into ferrous chloride and sulphide of iron. The vapors on passing out of B through the tube, which is kept at about 300°, deposit spangles of ferrous chloride, which is without sensible tension at that temperature. The vapors then enter D, a cast-iron cylinder in which are three cast-iron boats each containing 300 grms. of sodium. It is sufficient to heat this cylinder barely to a dull-red heat in its lower part, for the reaction once commenced disengages enough heat to complete itself, and it is often necessary to take away all the fire from it. There is at first produced in the first boat some aluminium and some sodium chloride, which latter combines with the excess of Al^2Cl^6 to form the volatile chloride $Al^2Cl^6.2NaCl$. These vapors of double chloride condense on the second boat and are decomposed by the sodium to aluminium and sodium chloride. A similar reaction takes place in the third boat when all the sodium of the second has disappeared. When on raising the cover it is seen that the reactions are over, the boats are taken out, immediately replaced by others, and are allowed to cool covered by empty boats. In this first operation the reaction is rarely complete, for the sodium is protected by the layer of NaCl formed at its expense. To make this disappear, the contents of the boats are put into cast iron pots or earthen crucibles, which are heated until the Al^2Cl^6 begins to volatilize. Then the pots or crucibles are cooled and there is taken from the upper part of their contents a layer of NaCl, almost pure, while underneath are found globules of aluminium, which are separated from the residue by washing with water. Unfortunately, the water in dissolving the Al^2Cl^6 of the flux exercises on the metal a very rapid destructive action, and only the globules larger than the head of a pin are saved from this washing. These are gathered together, dried, melted in an earthen crucible, and pressed together with a clay rod. The button is then cast in an ingot mould. It is important in this operation to employ only well purified sodium, and not to melt the aluminium if it still contains any sodium, for in this case the metal takes fire and burns as long as any of the alkaline metal remains in it. In such a case it is necessary to remelt in presence of a little $Al^2Cl^6.2NaCl$."

Deville says later, "Such was the detestable process by means of which we made the ingots of aluminium which were sent to the Exposition."

Deville, after this, tried some experiments in which he used sodium vapor, and he thus reports his results in his book: "This process, which I have not perfected, is very easy to operate, and gave me very pure metal at the first attempt. I operate as follows: I fill a mercury bottle with a mixture of chalk, carbon, and carbonate of soda, in the proportions best for generating sodium. An iron tube about ten centimetres long is screwed to the bottle, and the whole placed in a wind furnace, so that the bottle is heated to red-white and the tube is red to its end. The end of the tube is then introduced into a hole made in a large earthen crucible about one-fourth way from the bottom, so that the end of the tube just reaches the inside surface of the crucible. The carbonic oxide (CO) disengaged burns in the bottom of the crucible, heating and drying it; afterwards the sodium flame appears, and then pieces of Al^2Cl^6 are thrown into the crucible from time to time. The salt volatilizes and decomposes before this sort of tuyere from which issues the reducing vapor. Al^2Cl^6 is added when the vapors coming from the crucible cease to be acid, and when the flame of sodium burning in the atmosphere of Al^2Cl^6 loses its brightness. When the operation is finished, the crucible is broken and there is taken from the walls below the entrance of the tube a saline mass composed of NaCl, a considerable quantity of globules of aluminium, and some sodium carbonate, which latter is in larger quantity the slower the operation was performed. The globules are detached by plunging the saline mass into water, when it becomes necessary to notice the reaction of the water on litmus. If the water becomes acid, it is renewed often; if alkaline, the mass impregnated with metal must be digested in nitric acid diluted with three or four volumes of water, and so the metal is left intact. The globules are reunited by melting with the precautions before given."

Deville modified these methods in various ways. Al^2Cl^6 is a deliquescent salt, difficult of preservation, and so was soon replaced by $Al^2Cl^6.2NaCl$, which does not present these inconveniences. The double chloride, however, does draw some moisture and holds it energetically, from which it results that at a high temperature it will give rise to some alumina, which encloses the globules of metal with a thin coating and so hinders their easy reunion into a button. Deville remarked that the presence of fluorides facilitated the reunion of these globules, which he attributed to their dissolving the coat of Al^2O^3 on them. Since then, the employment of a fluoride as a flux is considered necessary to overcome the effect produced primarily by the $Al^2Cl^6.$-$2NaCl$ holding moisture so energetically. The first fluoride employed by Deville was fluorspar, which was soon replaced by cryolite. This opens up the subject of the reduction of aluminium from cryolite, and since Percy and Rose both preceded Deville in using it, I will first give their investigations, following with those which Deville published in 1859.

Reduction from Cryolite.

We will here give H. Rose's entire paper, as an account of this eminent chemist's investigations written out by himself with great detail, describing failures as well as successes, cannot but be of value to all interested in the production of aluminium.*

"Since the discovery of aluminium by Wöhler, Deville has recently devised the means of procuring the metal in large, solid masses, in which condition it exhibits properties with which we were previously unacquainted in its more pulverulent form as procured by Wöhler's method. While, for instance, in the latter state it burns vividly to white earthy alumina on being ignited, the fused globules may be heated to redness without perceptibly oxidizing. These differences may be ascribed to the greater amount of division on the one hand and of density on the other. According to Deville, however, Wöhler's metal contains platinum, by which he explains its difficulty of fusion, although it affords white alumina by combustion. Upon the publication of Deville's researches I also tried to produce aluminium by the decomposition of $Al^2Cl^6.2NaCl$ by means of sodium. I did not, however, obtain satisfactory results. Moreover, Prof. Rammelsberg, who followed exactly the

* Pogg. Annalen, Sept. 1855.

method of Deville, obtained but a very small product, and found it very difficult to prevent the cracking of the glass-tube in which the experiment was conducted by the action of the vapor of sodium on Al^2Cl^6. It appeared to me that a great amount of time, trouble, and expense, as well as long practice, was necessary to obtain even small quantities of this remarkable metal.

"The employment of Al^2Cl^6 and its compounds with alkali chlorides is particularly inconvenient, owing to their volatility, deliquescence, and to the necessity of preventing all access of air during their treatment with sodium. It very soon occurred to me that it would be better to use the fluoride of aluminium instead of the chloride; or rather the combination of the fluoride with alkaline fluorides, such as we know them through the investigations of Berzelius, who pointed out the strong affinity of Al^2F^6 for NaF and KF, and that the mineral occurring in nature under the name of Cryolite was a pure compound of Al^2F^6 and NaF.

"This compound is as well fitted for the preparation of aluminium by means of sodium as Al^2Cl^6 or $Al^2Cl^6.2NaCl$. Moreover, as cryolite is not volatile, is readily reduced to the most minute state of division, is free from water and does not attract moisture from the air, it affords peculiar advantages over the above-mentioned compounds. In fact, I succeeded with much less trouble in preparing aluminium by exposing cryolite together with sodium to a strong red heat in an iron crucible, than by using Al^2Cl^6 and its compounds. But the scarcity of cryolite prevented my pursuing the experiments. In consequence of receiving, however, from Prof. Krantz, of Bonn, a considerable quantity of the purest cryolite at a very moderate price ($2 per kilo), I was enabled to renew the investigation.

"I was particularly stimulated by finding, most unexpectedly, that cryolite was to be obtained here in Berlin commercially at an inconceivably low price. Prof. Krantz had already informed me that cryolite occurred in commerce in bulk, but could not learn where. Shortly after, M. Rudel, the manager of the chemical works of H. Kunheim, gave me a sample of a coarse white powder, large quantities of which were brought from Greenland, by way of Copenhagen, to Stettin, under the name of mineral soda, and at the price of $3 per centner. Samples had been sent to the soap boilers, and a soda-lye had been extracted from it by means of quicklime, especially adapted to the preparation of many kinds of soap, probably from its containing alumina. It is a fact, that powdered cryolite is completely decomposed by quicklime and water. The fluoride of lime formed contains no alumina, which is all dissolved by the caustic soda solution; and this, on its side, is free from fluorine, or only contains a minute trace. I found this powder to be of equal purity to that received from Prof.

Krantz. It dissolved without residue in HCl (in platinum vessels); the solution evaporated to dryness with H^2SO^4, and heated till excess of acid was dissipated, gave a residue which dissolved completely in water, with the aid of a little HCl. From this solution, ammonia precipitated a considerable quantity of alumina. The solution filtered from the precipitate furnished, on evaporation, a residue of sulphate of soda, free from potash. Moreover, the powder gave the well-known reactions of fluorine in a marked degree. This powder was cryolite of great purity: therefore the coarse powder I first obtained was not the form in which it was originally produced. It is now obtainable in Berlin in great masses; for the preparation of aluminium it must, however, be reduced to a very fine powder.

"In my experiments on the preparation of aluminium, which were performed in company with M. Weber, and with his most zealous assistance, I made use of small iron crucibles, $1\frac{3}{4}$ inches high and $1\frac{3}{8}$ inches upper diameter, which I had cast here. In these I placed the finely divided cryolite between thin layers of sodium, pressed it down tight, covered with a good layer of potassium chloride (KCl), and closed the crucible with a well-fitting porcelain cover. I found KCl the most advantageous flux to employ; it has the lowest specific gravity of any which could be used, an important point when the slight density of the metal is taken into consideration. It also increases the fusibility of the sodium fluoride. I usually employed equal weights of cryolite and KCl, and for every five parts of cryolite two parts of sodium. The most fitting quantity for the crucible was found to be ten grammes of powdered cryolite. The whole was raised to a strong red heat by means of a gas-air blowpipe. It was found most advantageous to maintain the heat for about half an hour, and not longer, the crucible being kept closely covered the whole time; the contents were then found to be well fused. When quite cold the melted mass is removed from the crucible by means of a spatula, this is facilitated by striking the outside with a hammer. The crucible may be employed several times, at last it is broken by the hammer blows. The melted mass is treated with water, when, at times only, a very minute evolution of hydrogen gas is observed, which has the same unpleasant odor as the gas evolved during solution of iron in HCl. The carbon contained in this gas is derived from a very slight trace of naphtha adhering to the sodium after drying it. On account of the difficult solubility of NaF, the mass is very slowly acted on by water, although the insolubility is somewhat diminished by the presence of the KCl. After twelve hours the mass is softened so far that it may be removed from the liquid and broken down in a porcelain mortar. Large globules of aluminium are then discovered, weighing from 0.3 to 0.4 or even 0.5 gramme, which may be separated out. The smaller globules cannot well be separated from the undecomposed cryolite and the alumina always produced by washing, owing to their being specifically lighter than the latter. The whole is treated with HNO^3 in the cold. The Al^2O^3 is not dissolved thereby, but the little globules then first assume their true metallic lustre. They are dried and rubbed on fine silk muslin; the finely-powdered, undecomposed cryolite and Al^2O^3 pass through, while the globules remain on the gauze. The mass should be treated in a platinum or silver vessel, a porcelain vessel would be powerfully acted on by the NaF. The solution, after standing till clear, may be evaporated to dryness in a platinum capsule, in order to obtain the NaF, mixed, however, with much KCl. The small globules may be united by fusion in a small, well-covered, porcelain crucible, under a layer of KCl. They cannot be united without a flux. They cannot be united by mere fusion, like globules of silver, for instance, for, though they do not appear to oxidize on ignition in the air, yet they become coated with a scarcely perceptible film of oxide, which prevents their running together into a mass. This fusion with KCl is always attended with loss of aluminium. Buttons weighing 0.85 grm. lost, when so treated, 0.05 grm. The KCl when dissolved in water left a small quantity of Al^2O^3 undissolved, but the solution contained none. Another portion of the metal had undoubtedly decomposed the KCl; and a portion of the Al^2Cl^6 and KCl must have been volatilized during fusion (other metals, as copper and silver, behave in a similar manner—Pogg. lxviii. 287). I therefore followed the instructions of Deville, and melted the globules under a stratum of $Al^2Cl^6.2NaCl$ in a covered porcelain crucible. The salt was melted first, and then the globules of metal added to the melted mass. There is no loss, or a very trifling one of a few milligrammes of metal, by this proceeding. When the aluminium is fused under KCl its surface is not perfectly smooth, but exhibits minute concavities; with $Al^2Cl^6.2NaCl$ this is not the case. The readiest method of preparing the $Al^2Cl^6.2NaCl$ for this purpose is by placing a mixture of alumina and carbon in a glass tube, as wide as possible, and inside this a tube of less diameter, open at both ends, and containing NaCl. If the spot where the mixture is placed be very strongly heated, and that where the NaCl is situated, more moderately, while a current of chlorine is passed through the tube, the vapor of Al^2Cl^6 is so eagerly absorbed by the NaCl that no Al^2Cl^6, or, at most, a trace, is deposited in any other part of the tube. If the smaller tube be weighed before the operation, the amount absorbed is readily determined. It is not uniformly combined with the NaCl, for that part which is nearest to the mixture of charcoal and alumina will be found to have absorbed the most.

"I have varied in many ways the process for the preparation of aluminium, but in the end have returned to the one just described. I often placed the sodium in the bottom of the crucible, the powdered cryolite about it, and the KCl above all. On proceeding in this manner, it was observed that much sodium was volatilized, burning with a strong, yellow flame, which never occurred when it was cut into thin slices and placed in alternate layers with the cryolite, in which case the process goes on quietly. When the crucible begins to get red hot, the temperature suddenly rises, owing to the commencement of the decomposition of the compound; no lowering of the temperature should be allowed, but the heat should be steadily maintained, not longer, however, than half an hour. By prolonging the process a loss would be sustained, owing to the action of the KCl on the aluminium. Nor does the size of the globules increase on extending the time even to two hours; this effect can only be produced by obtaining the highest possible temperature. If the process be stopped, however, after five or ten minutes of very strong heat, the production is very small, as the metal has not had sufficient time to conglomerate into globules, but is in a pulverulent form and burns to Al^2O^3 during the cooling of the crucible. No advantage is gained by mixing the cryolite with a portion of chloride before placing it between the layers of sodium, neither did I increase the production by using $Al^2Cl^6.2NaCl$ to cover the mixture instead of KCl. I repeatedly employed NaCl, decrepitated, as a flux in the absence of KCl, without remarking any important difference in the amount of metal produced, although a higher temperature is in this case required. The operations may also be conducted in refractory unglazed crucibles made of stoneware, and of the same dimensions, although they do not resist so well the action of the sodium fluoride at any high heats, but fuse in one or more places. The iron crucibles fuse, however, when exposed to a very high temperature in a charcoal fire. The product of metal was found to vary very much, even when operating exactly in the manner recommended and with the same quantities of materials. I never succeeded in reducing the whole amount of metal contained in the cryolite (which contains only 13 per cent. Al). By operating on 10 grammes of cryolite, the quantity I always employed in the small Fe crucible, the most successful result was 0.8 grm. But 0.6 or even 0.4 grm. may be considered favorable; many times I obtained only 0.3 grm., or even less. These very different results depend on various causes, more particularly, however, on the degree of heat obtained. The greater the heat the greater the amount of large globules, and the less amount of minutely divided metal to oxidize during the cooling of the crucible. I succeeded once or twice in reducing nearly the whole of the metal to one single button, weighing 0.5 grm., at a very high heat in a stoneware crucible. I could not always obtain the same heat with the blowpipe, as it depended in some degree on the pressure in the gasometer in the gasworks, which varies at different hours of the day. The following experiment will show how great the loss of metal may be owing to oxidation during the slow cooling of the crucible and its contents: In a large iron crucible were placed 35 grms. of cryolite in alternate layers with 14 grms. of sodium and the whole covered with a thick stratum of KCl. The crucible, covered by a porcelain cover, was placed in a larger earthen one, also covered, and the whole exposed to a good heat in a draft furnace for one hour and cooled as slowly as possible. The product in this case was remarkably small, for 0.135 grm. of aluminium was all that could be obtained in globules. The differences in the amounts reduced depend also in some degree on the more or less successful stratification of the sodium with the powered cryolite, as much of the latter sometimes escapes decomposition. The greater the amount of sodium employed, the less likely is this to be the case; however, owing to the great difference in their prices, I never employed more than 4 grms. of sodium to 10 grms. of cryolite. In order to avoid this loss by oxidation I tried another method of preparation: Twenty grms. of cryolite were heated intensely in a gun-barrel in a current of hydrogen, and then the vapor of 8 grms. of sodium passed over it. This was effected simply by placing the sodium in a little iron tray in a part of the gun-barrel without the fire, and pushing it forward when the cryolite had attained a maximum temperature. The operation went on very well, the whole being allowed to cool in a current of hydrogen. After the treatment with water, in which the sodium fluoride dissolved very slowly, I obtained a black powder, consisting for the most part of iron. Its solution in HCl gave small evidence of Al. The small amounts I obtained, however, should not deter others from making these experiments. These are the results of first experiments on which I have not been able to expend much time. Now that cryolite can be procured at so moderate a price, and sodium, by Deville's improvements, will in future become so much cheaper, it is in the power of every chemist to engage in the preparation of aluminium, and I have no doubt that in a short time methods will be found affording a much more profitable result.

"For the rest, I am of opinion that cryolite is the best adapted of all the compounds of aluminium for the preparation of this metal. It deserves the preference over $Al^2Cl^6.2NaCl$ or Al^2Cl^6, and it might still be employed with great advantage even if its price were to rise considerably. The attempts at preparing aluminium direct from Al^2O^3 have as yet been unattended with success. Potassium and sodium appear only to reduce metallic oxides when the potash and soda produced are capable of forming compounds with a portion of the oxide remaining as such. Pure potash and soda, with whose properties we are very slightly acquainted, do not appear to be formed in this case. Since, however, alumina combines so readily with the alkalies to form aluminates, one would be inclined to believe that the reduction of Al^2O^3 by the alkali metals should succeed. But even were it possible to obtain the metal directly from Al^2O^3, it is very probable that cryolite would long be preferred should it remain at a moderate price, for it is furnished by nature in a rare state of purity, and the aluminium is combined in it with sodium and fluorine only, which exercise no prejudicial influence on the properties of the metal, whereas Al^2O^3 is rarely found in nature in a pure state and in a dense, compact condition, and to prepare Al^2O^3 on a large scale, freeing it from those substances which would act injuriously on the properties of the metal, would be attended with great difficulty.

"The buttons of aluminium which I have prepared are so malleable that they may be beaten and rolled out into the finest foil without cracking on the edges. They have a strong metallic lustre. Some small pieces, not globular, however, were found in the bottom of the crucible, and occasionally adhering to it, which cracked on being hammered, and were different in color and lustre from the others. They were evidently not so pure as the greater number of the globules, and contained iron. On sawing through a large button weighing 3.8 grms., it could readily be observed that the metal for about half a line from the exterior was brittle, while in the interior it was soft and malleable. Sometimes the interior of a globule contained cavities. With Deville, I have occasionally observed aluminium crystallized. A large button became striated and crystalline on cooling. Deville believes he has observed regular octahedra, but does not state this positively. According to my brother's examination, the crystals do not belong to any of the regular forms. As I chanced on one occasion to attempt the fusion of a large, flattened-out button of rather impure aluminium, without a flux, I observed, before the heat was sufficient to fuse the mass, small globules sweating out from the surface. The impure metal being less fusible than pure metal, the latter expands in fusing and comes to the surface."

Such were the results given to the world by H. Rose. After their publication, many minds were turned towards this field, and it was discovered that some six months previously Dr. Percy had accomplished the same results, and had even shown them to the Royal Institution, but with the singular fact of exciting very little attention. These facts are stated at length in the following paper, written by Allan Dick, Esq., which appeared in November, 1855, two months after the publication of H. Rose's paper:—*

"In the last number of this magazine was the translation of a paper by H. Rose, of Berlin, describing a method of preparing aluminium from cryolite. Previously, at the suggestion of Dr. Percy, I had made some experiments on the same subject in the metallurgical laboratory of the School of Mines, and as the results obtained agree very closely with those of Mr. Rose, it may be interesting to give a short account of them now, though no detailed description was published at the time, a small piece of metal prepared from cryolite having simply been shown at the weekly meeting of the Royal Institution, March 30, 1855, accompanied by a few words of explanation by Faraday.

"Shortly after the publication of Mr. Deville's process for preparing aluminium from Al^2Cl^6, I tried, along with Mr. Smith, to make a specimen of the metal, but we found it much more difficult to do than Deville's paper had led us to anticipate, and had to remain contented with a much smaller piece of metal than we had hoped to obtain. It is, however, undoubtedly only a matter of time, skill, and expense to join successful practice with the details given by Deville. Whilst making

* Phil. Mag., Nov. 1855.

these experiments Dr. Percy had often requested us to try whether cryolite could be used instead of the chlorides, but some time elapsed before we could obtain a specimen of the mineral. The first experiments were made in glass tubes sealed at one end, into which alternate layers of finely powdered cryolite and sodium cut into small pieces were introduced, and covered in some instances with a layer of cryolite, in others by NaCl. The tube was then heated over a gas blowpipe for a few minutes till decomposition had taken place and the product was melted. When cold, on breaking the tube, it was found that the mass was full of small globules of aluminium, but owing to the specific gravity of the metal and flux being nearly alike, the globules had not collected into a button at the bottom. To effect this, long continued heat would be required, which cannot be given in glass tubes owing to the powerful action of the melted fluoride on them. To obviate this difficulty a platinum crucible was lined with magnesia by ramming it in hard, and subsequently cutting out all but a lining. In this, alternate layers of cryolite and sodium were placed, with a thickish layer of cryolite on top. The crucible was covered with a tight-fitting lid, and heated to redness for about half an hour over a gas blowpipe. When cold it was placed in water, and after soaking for some time the contents were dug out, gently crushed in a mortar, and washed by decantation. Two or three globules of aluminium, tolerably large considering the size of the experiment, were obtained, along with a large number of very small ones. The larger ones were melted together under KCl. Some experiments made in iron crucibles were not attended with the same success as those of Rose, no globules of any considerable size remained in the melted fluorides; the metal seemed to alloy on the sides of the crucible, which acquired a color like zinc. It is possible that this difference may have arisen from using a higher temperature than Rose, as we made these experiments in a furnace, not over the blowpipe. Porcelain and clay crucibles were also tried, but laid aside after a few experiments, owing to the action of the fluorides upon them, which in most cases was sufficient to perforate them completely."

The above papers, Rose's and Dick's, contain all the published researches with cryolite until Deville's attention was turned towards it. He then took up the subject with his accustomed thoroughness. The following pages are taken from his 'De l'Aluminium,' the subject not being given in its entirety, but only the most important points. He published the first account of these researches in 'Ann. de Chem. et de Phys.' [3], xlvi. 451:—

"I have repeated and confirmed all the experiments of Dr. Percy and H. Rose, using the specimens of cryolite which I obtained from London through the kindness of MM. Rose and Hofmann. I have, furthermore, reduced cryolite mixed with NaCl by the battery, and I believe that this will be an excellent method of covering with aluminium all the other metals, copper in particular. Anyhow, its fusibility is considerably increased by mixing it with $Al^2Cl^6.2KCl$. Cryolite is a double fluoride of aluminium and sodium, containing 13 per cent. Al, 32.5 per cent. Na, and 54.5 per cent. F. Its formula is $Al^2F^6.6NaF$. I have verified these facts myself by many analyses."

Deville then gives a description of methods of making cryolite artificially, which is unnecessary to repeat here, for natural cryolite is so cheap that these methods are of no practical importance. He continues:—

"In reducing the cryolite I placed the finely-pulverized mixture of cryolite and NaCl in alternate layers with sodium in a porcelain crucible. The uppermost layer is of pure cryolite, covered with NaCl. The mixture is heated just to complete fusion, and, after stirring with a pipe-stem, is let cool. On breaking the crucible, the aluminium is often found united in large globules easy to separate from the mass. The metal always contains silicon, which increases the depth of its natural blue tint and hinders the whitening of the metal by nitric acid, because of the insolubility of the silicon in that acid. M. Rose's metal is very ferruginous. I have verified all M. Rose's observations, and I agree with him concerning the return of metal, which I have always found very small. There are always produced in these operations brilliant flames, which are observed in the scoria floating on the aluminium, and which are due to gas burning and exhaling a very marked odor of phosphorus. In fact, P^2O^5 exists in cryolite, as one may find by treating a solution of the mineral in sulphuric acid with molybdate of ammonia, according to H. Rose's reaction.

"The facility with which aluminium unites in fluorides is due without doubt to the property which these possess of dissolving the alumina on the surface of the globules at the moment of their formation, and which the sodium is unable to reduce. I had experienced great difficulty by obtaining small quantities of metal poorly united, when I reduced the $Al^2Cl^6.2NaCl$ by sodium; M. Rammelsberg, who often made the same attempts, tells me he has had a like experience. But, I am assured by a scrupulous analysis that the quantity of metal reduced by the sodium is exactly that which theory indicates, although after many operations there is found only a gray powder, resolving itself under the microscope into a multitude of small globules. The fact is simply that $Al^2Cl^6.2NaCl$ is a very poor flux for aluminium. MM. Morin, Debray, and myself have undertaken to correct this bad effect by the introduction of a solvent for the Al^2O^3 into the saline slag which accompanies the aluminium at the moment of its formation. At first, we found it an improvement to condense the vapors of Al^2Cl^6, previously purified by iron, directly in NaCl, placed for this purpose in a crucible and kept at a red heat. We produced in this way, from highly colored material, a double chloride very white and free from moisture, and furnishing on reduction a metal of fine appearance. We then introduced fluorspar (CaF^2) into the composition of the mixture to be reduced, and we obtained good results with the following proportions:—

$Al^2Cl^6.2NaCl$	400 grammes.
NaCl	200 "
CaF^2	200 "
Na	75 to 80 "

The double chloride ought to be melted and heated almost to low red heat at the moment it is employed, the NaCl calcined and at a red heat or melted, and the CaF^2 pulverized and well dried. The double chloride, NaCl and CaF^2 are mixed and alternated in layers in the crucible with sodium. The top layer is of the mixture, and the cover is NaCl. Heat gently, at first, until the reaction ends, and then to a heat about sufficient to melt silver. The crucible, or at least that part of it which contains the mixture, ought to be of a uniform red tint, and the material perfectly liquid. It is stirred a long time and cast on a well-dried, chalked plate. There flows out first a very limpid liquid, colorless and very fluid, then a gray material, a little more pasty, which contains aluminium in little grains, and is set aside, and finally a button with small, metallic masses which of themselves ought to weigh 20 grms. if the operation has succeeded well. On pulverizing and sieving the gray slag, 5 or 6 grms. of small globules are obtained, which may be pressed together by an earthen rod in an ordinary crucible heated to redness. The globules are thus reunited, and when a sufficient quantity is collected the metal is cast into ingots. In a well-conducted operation, 75 grms. of Na ought to give a button of 20 grms. and 5 grms. in grains, making a return of one Al from three of Na. Theory indicates one to two and a half, or 30 grms. of Al from 75 of Na. But all the efforts which have been made to recover from the insoluble slag the 4 or 5 grms. of metal not united but easily visible with a glass, have been so far unsuccessful. There is, without doubt, a knack, a particular manipulation on which depends the success of an operation which would render the theoretical amount of metal, but we lack it yet. These operations take place, in general, with more facility on a large scale, so that we may consider fluorspar as being suitable for serving in the manufacture of aluminium in crucibles. We employed very pure fluorspar, and our metal was quite exempt from silicon. It is true that we took a precaution which is necessary to adopt in operations of this kind; we plastered our crucibles inside with a layer of aluminous paste, the composition of which has been given in ' Ann. de Chem. et de Phys.,' xlvi.

195. This paste is made of calcined alumina and an aluminate of lime, the latter obtained by heating together equal parts of chalk and alumina to a high heat. By taking about four parts calcined alumina and one of aluminate of lime well pulverized and sieved, moistening with a little water, there is obtained a paste with which the inside of 'an earthen crucible is quickly and easily coated. The paste is spread evenly with a porcelain spatula, and compressed strongly until its surface has become well polished. It is allowed to dry, and then heated to bright redness to season the coating, which does not melt, and protects the crucible completely against the action of the aluminium and fluorspar. A crucible will serve several times in succession provided that the new material is put in as soon as the previous charge is cast. The advantages of doing this are that the mixture and the sodium are put into a crucible already heated up, and so lose less by volatilization because the heating is done more quickly, and the crucible is drier than if a new one had been used or than if it had been let cool. A new crucible should be heated to at least 300° or 400° before being used. The saline slag contains a large quantity of calcium chloride, which can be washed away by water, and an insoluble material from which aluminium fluoride can be volatilized.

" Yet the operation just described, which was a great improvement on previous ones, requires many precautions and a certain skill of manipulation to succeed every time. But, nothing is more easy or simple than to substitute cryolite for the fluorspar. Then the operation is much easier. The amount of metal produced is not much larger, although the button often weighs 22 grammes, yet if cryolite can only be obtained in abundance in a continuous supply, the process which I will describe will become most economical. The charge is made up as before, except introducing cryolite for CaF^2. In one of our operations we obtained, with 76 grms. of sodium, a button weighing 22 grms. and 4 grms. in globules, giving a yield of one Al to two and eight-tenths parts sodium, which is very near to that indicated by theory. The metal obtained was of excellent quality. However, it contained a little iron coming from the Al^2Cl^6, which had not been purified perfectly. But iron does not injure the properties of the metal as copper does; and, save a little bluish coloration, it does not alter its appearance or its resistance to physical and chemical agencies.

"Process with cryolite alone: The process adopted in the works at Amfreville, near Rouen, directed by Tissier Bros., is the same as that described by Percy and Rose. The details which I give are taken from MM. Tissier's own account of their process." (Deville then gives the details of the process outlined by Rose (see p. 103), of reducing in iron crucibles; which it is not necessary to repeat.)

"I obtained a good specimen of commercial aluminium thus extracted from cryolite; and M. Demondèur has been so kind as to make an analysis of it, with the following results: Si 4.4; Fe 0.8; Al 94.8.

" M. Rose has recommended iron vessels for this operation, because of the rapidity with which alkaline fluorides attack earthen crucibles and so introduce considerable silicon into the metal. Unfortunately, these iron crucibles introduce iron into the metal. This is an evil inherent to this method, at least in the present state of the industry. The inconveniences of this method result in part from the high temperature required to complete the operation, and from the crucible being in direct contact with the fire, by which its sides are heated hotter than the metal in the crucible. The metal itself, placed in the lower part of the fire, is hotter than the slag. This, according to my observations, is an essentially injurious condition. The slag ought to be cool, the metal still less heated, and the sides of the vessel where the fusion occurs ought to be as cold as possible. The yield from cryolite, according to Rose's and my own observations, is also very small. M. Rose obtained from 10 of cryolite and 4 of Na about 0.5 of Al. This is due to the affinity of aluminium for fluorine, which must be very strong not only with relation to its affinity for sodium but even for calcium, and this affinity appears to increase with the temperature, as was found in my laboratory. Cryolite is convenient to employ as a flux to add to the mixture which is fused, especially when operating on a small scale;

but it is fortunate that it is not indispensable, for no one would wish to establish an industry on the employment of a material which is of uncertain supply."

We here close what Deville has written on the use of cryolite. The process was that used by Tissier Bros. at Rouen, but was finally abandoned there and the works closed. We find a little improvement on Deville's process suggested by Wöhler,[*] in which he shows how to perform the reduction in an earthen crucible. The finely pulverized cryolite is mixed with an equal weight of a flux containing 7 NaCl to 9 KCl. This mixture is then placed in alternate layers with sodium in the crucible, 50 parts of the mixture to 10 of sodium, and heated gradually just to its fusing point. The metal thus obtained is free from silicon, but only one-third of the aluminium in the cryolite is obtained. In spite of the small yield, this method was used for some time by Tissier Bros. Cryolite has also been treated at Nanterre, by a different process, but the aluminium produced contained phosphorus. So, while the exclusive use of cryolite in the preparation of aluminium is now renounced, it has retained the office of a flux.

Watts gives the following paragraph in connection with the reduction of cryolite: " A peculiar apparatus for effecting the reduction of aluminium,

---

[*] Ann. der Chem. und Pharm. 99, 255.

either from $Al^2Cl^6.2NaCl$ or from cryolite, the object of which is to prevent loss of sodium by ignition, has been invented and patented by W. F. Gerhard.* It consists of a reverberatory furnace having two hearths, or of two crucibles, or of two reverberatory furnaces, placed one above the other and communicating by an iron pipe. In the lower is placed a mixture of sodium with the aluminium compound, and in the upper a stratum of NaCl, or of a mixture of NaCl and cryolite, or of the slag obtained in a previous operation. This charge, when melted, is made to run into the lower furnace in quantity sufficient to completely cover the mixture contained therein, and so to protect it from the air. The mixture thus covered is reduced as by the usual operation."

Watts thus summarizes the use of cryolite: " The chief inducement for using it as a source of aluminium is that it is a natural product obtained with tolerable facility, and enables the manufacturer to dispense with the troublesome and costly preparation of $Al^2Cl^6.2NaCl$. But the metal thus obtained is less pure than that obtained by other processes. If earthenware crucibles are used, the metal is contaminated with silicon, because the sodium fluoride produced acts strongly on the siliceous matter of the crucible, while if an iron crucible be used, the metal takes up some iron.

* Eng. Pat. 1858, No. 2247.

The best use of cryolite is as a flux in the preparation of aluminium from $Al^2Cl^6.2NaCl$, in which case the slag is not sodium fluoride but aluminium fluoride, which acts but slightly on the containing vessel."

General Remarks.

I have now given the metallurgy of aluminium through what may be called its experimental stage up to its practical industrial manufacture. Up to this period, which I will place at about 1859, the object has been to produce the metal at any cost, only produce it. "To learn how" engrossed the attention of the investigators, who troubled themselves very little about the ultimate cost. They must learn first how to do the thing and afterwards devote their energies to cheapening the process discovered. But, in 1859, the works at Amfreville, near Rouen, under the direction of the Tissiers, is producing aluminium from cryolite; Morin & Co., at Nanterre, are making it, though not in such large quantities as Tissier, but they soon after move to Salindres, and set up so large a plant that a year or so afterwards the Tissiers were driven from the business. Such is then the state of the industry. We find that in the next fifteen or twenty years very little advance is chronicled. At Salindres, the processes given by Deville were used somewhat improved and perfected, but yet the same processes. It is only within the last ten years that any improvements of a radical nature, such as Webster's, Frishmuth's, and Cowles, have been brought into the industry.

So, from now on we will treat the subject in the order usually adopted in presenting it; *i. e.*, first give a short sketch of the metallurgy of sodium up to the present time, then a review of the manufacture of alumina and its conversion into $Al^2Cl^6.2NaCl$, ending with a full description of the process as now carried on at Salindres, and a few attempts which have been made to improve it. Afterwards, leaving the old Deville process and its improvements, I will give as full an account as I have been able to gather of the various methods proposed to produce aluminium without the use of sodium.

PART VI.

THE MANUFACTURE OF SODIUM.

As already observed, we will not go extensively into the metallurgy of this metal. Some years ago, in order to treat fully of the metallurgy of aluminium, it would have been as necessary to accompany it with all the details of the manufacture of sodium as to give the details of the reduction of the aluminium, because the manufacture of the former was carried on solely in connection with that of the latter. But now sodium has come out of the list of chemical curiosities and has become an article of commerce, used for many other purposes than the reduction of aluminium, though that is still its chief use. So we regard the manufacture of sodium as a separate metallurgical subject, still intimately connected with that of aluminium, but yet so far distinct from it as to deserve a metallurgical treatise of its own. Moreover, the metallurgy of sodium is very much as Deville left it, it has been very little improved since then, and so almost all the details of its manufacture are to be found in English in any good book on chemistry. To such works I refer the reader for fuller accounts than are given here. The following summary is taken principally from Mierzinski.

Sodium was first isolated by Davy by the use of electricity in the year 1808.* Later, Gay Lussac and Thenard made it by decomposing at a very high temperature a mixture of Na^2CO^3 and iron filings.† On April 30, 1808, Curaudau announced that he had succeeded in producing potassium or sodium without using iron, simply by decomposing K^2CO^3 or Na^2CO^3 by means of animal charcoal. Brünner continuing this investigation used instead of animal charcoal the so-called black flux, the product obtained by calcining crude tartar from wine barrels. He was the first to use the wrought-iron mercury bottles. The mixture was heated white hot in a furnace, the sodium volatilized and was condensed in an iron tube screwed into the top of the flask, which projected from the furnace and was cooled with water. In Brünner's experiments he only obtained three per cent. of the weight of the mixture as metallic sodium, the rest of the metal being lost as vapor.

Donny and Mareska gave the condenser the form which with a few modifications it retains to-day. It was of iron, 4 millimetres thick, and was made in the shape of a book, having a length of about 100 centimetres, breadth 50, and depth 6 (see Fig. 2). This form is now so well known that a further

---

* Phil. Trans., 1808.
† Recherches Physico-chemiques, 1810.

description is unnecessary. With this condenser the greatest difficulty of the process was removed, and the operation could be carried on in safety.

Fig. 2.

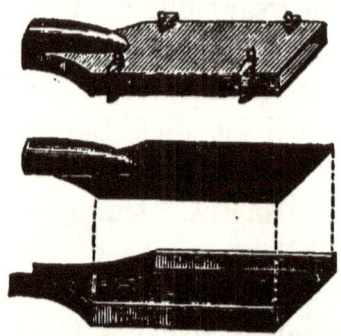

This apparatus was devised and used by Donny and Mareska in 1854, with the supervision of Deville, and the whole process as used by them is the same that the Tissier Bros. took with them and operated at their works at Rouen, and their description accords with that given by Deville, which is as follows:—

The Na^2CO^3 is first well dried at a high temperature, then mixed with well dried pulverized charcoal and chalk, ground to the finest powder, the success of the operation depending on the fineness of this mixture. The proportions of these to use is various. One simple mixture is of

Na^2CO^3	30
Coal	13
Chalk	5
Coke	5

THE MANUFACTURE OF SODIUM.

Deville recommends taking—

Na^2CO^3	1000
Coal	450
Chalk	175

The addition of chalk has the object of making the mixture less fusible and more porous, but has the disadvantage that the residue remaining in the retort after the operation is very impure, and it is impossible to add any of it to the succeeding charge; and also, some of it being reduced to caustic lime forms caustic alkali with some Na^2CO^3, which is then lost. When the mixture is well made it is subjected to a preliminary calcination. This is done in cast-iron cylinders, two of which are placed side by side in a furnace and heated to redness (see Fig. 3). This is continued till all the moisture,

Fig. 3.

carbonic acid, and any carburetted hydrogen from the coal cease coming off. The mass contracts, becomes white and somewhat dense, so that a larger amount of the mixture can now be treated in the retorts where the sodium is evolved. As soon as the outcoming gases burn with a yellow flame, showing sodium coming off, the calcination is stopped. The mixture is then immediately drawn out on to the stone floor of the shop, where it cools quickly and is then ready for the next operation. This calcination yields a mixture which without any previous reactions is just ready to evolve sodium when brought to the necessary temperature. This material is made into a sort of cylinder or cartridge and put into the decomposition retorts (see Fig. 4).

Fig. 4.

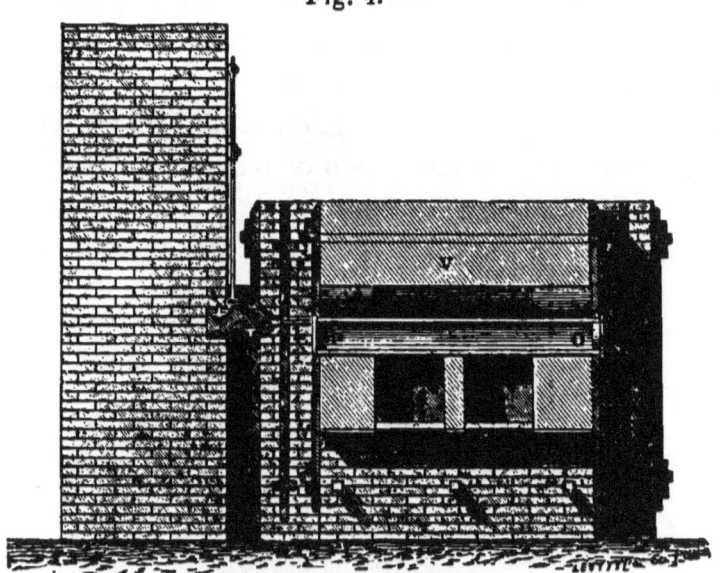

The charging should be done quickly. The final retorts are 120 centimetres long, 12 to 14 centimetres diameter, with walls 10 to 30 millimetres thick. These are of wrought iron, since cast iron would not stand the heat. At each end this retort is closed with wrought-iron stoppers and made tight with fire-clay. Through one stopper leads the pipe to the condenser, the other stopper is the one removed when the retort is to be recharged. These retorts are placed horizontally in rows in a furnace. Usually four are placed in a furnace, preferably heated by gas, such as the Siemens regenerative furnace or Bicheroux's, these being much more economical. In spite of all these precautions the retorts will be strongly attacked, and in order to protect them from the destructive action of a white heat for seven or eight hours they are coated with some kind of fire-proof material. The best for this purpose is graphite, which is made into cylinders inclosing the retorts, and which can remain in place till the furnace is worn out. These graphite cylinders not only protect the iron retorts, but prevent the diffusion of the gaseous products of the reaction into the hearth, and so support the retorts that their removal from the furnace is easily accomplished. Instead of these graphite cylinders the retorts may be painted with a mixture that melts at white heat and so enamels the outside. A mixture of alumina, sand, yellow earth, borax, and water-glass will serve very well in many cases. We would remark that the waste gases from this furnace can be used for the calcining of the mixture, or even for the reduction of the aluminium by sodium, where the manufacture of the former is connected with the making of the sodium. Donny and Mareska's condenser is the best to use.

As for the reduction of the sodium, the retort is first heated to redness, during which the stopper at the condenser end of the retort is left off. The charge is then rapidly put in, and the stopper at once put in place. The reaction begins almost at once and the operation is soon under full headway, the gases evolved burning from the upper slit of the condenser tube with a flame a foot long. The gases increase in volume as the operation continues, the flame becoming yellower from sodium and so intensely bright as to be insupportable to look at. Now has come the moment when the workman must quickly adapt the condenser to the condenser tube projecting from the retort, the joint being greased with tallow or paraffine. The sodium collects in this in a melted state and trickles out. The length of the operation varies, depending on the intensity of the heat and the quantity of the mixture; a charge may sometimes be driven over in two hours, and sometimes it takes eight. We can say, in general, that if the reaction goes on quickly a somewhat larger amount of sodium is obtained. The higher the heat used, however, the quicker the retorts are destroyed. The operation requires continual attention. From time to time, a workman with a prod opens up the neck of the

THE MANUFACTURE OF SODIUM. 137

condenser. But, if care is not taken the metal overflows: if this happens, the metal overflowing is thrown into some petroleum, while another man replaces the condenser with an empty one. The operation is ended when the evolution of gas ceases and the flame becomes short and feeble, while the connecting tube between the retort and condenser keeps clean and does not stop up. As soon as this occurs, the stopper at the charging end is removed, the charge raked out into an iron car, and a new charge being put in, the operation continues. After several operations the retorts must be well cleaned and scraped out. The sodium thus obtained is in melted bits or drops, mixed with carbon and Na^2CO^3. It must therefore be cleaned, which is done by melting it in a wrought-iron kettle under paraffin with a gentle heat, and then casting it into the desired shapes. The sodium is kept under a layer of oil or any hydrocarbon of high boiling point containing no oxygen.

Deville says that the temperature necessary for the reduction of sodium from Na^2CO^3 and carbon is not so high as is generally supposed. He says that it was M. Rivot's opinion that the retorts were not heated higher than the retorts at Veille-Montague used for reducing zinc. Tissier gives the reaction as

$$Na^2CO^3 + 2C = 3CO + 2Na.$$

The sodium is condensed, while the carbonic

oxide, carrying over some sodium, burns at the end of the apparatus. This would all be very simple if the reaction of carbonic oxide on sodium near the condensing point did not complicate matters, producing a black, infusible deposit of Na^2O and C, which on being melted always gives rise to a loss of sodium.

The foregoing is the process as perfected by Donny and Mareska, Deville, and Tissier. Only a few improvements have been made, the most important are the following:—

R. Wagner[*] uses paraffin in preference to paraffin oil in which to keep the sodium after making it. Only pure paraffin which has been melted a long time on a water bath and all its water driven off can be used. The sodium to be preserved is dipped in the paraffin melted on a water bath and kept at no higher heat than 55°, and the metal is thereby covered with a thick coat of paraffin which protects it from oxidation, and may then be put up in wooden or paper boxes. When the metal is to be used, it is easily freed from paraffin by simply warming it, since sodium melts at 95° to 96° C. and the paraffin at 50° to 60°.

The reduction of K^2CO^3 by carbon requires much less heat than that of Na^2CO^3, and, therefore, many attempts have been made to reduce potassium and sodium together, under circumstances where so-

---

[*] Dingler, 1883, p. 252.

dium alone would not be reduced. Dumas* added some K^2CO^3 to the regular sodium mixture; and separated the sodium and potassium from each other by a slow, tedious oxidation. R. Wagner† made a similar attempt. He says that not only does the reduction of both metals from a mixture of K^2CO^3, Na^2CO^3, and carbon work easier than Na^2CO^3 and carbon, but even caustic soda (NaOH) may be used with K^2CO^3 and carbon. Also, the melting point of potassium and sodium alloyed is much lower than that of either one alone, in consequence of which their boiling point and the temperature required for reduction are lower.

W. Weldon calculated the cost of sodium as seven to eight marks per kilo. The greater part of this is for retorts in which the operation takes place, and which are so quickly destroyed that the replacing of them forms half the cost of the metal. Compare with p. 172.

The latest announcement of advance in making sodium is from New York City, and is thus described in a New York paper:—‡

"When sodium was reduced in price to $1.50 per pound, it was thought to have touched a bottom figure, and all hope of making it any cheaper seemed fruitless. This cheapening was not brought about

* Handbuch der Angewandten Chemie, 1830, ii. 345.
† Dingler, 143, 343.
‡ New York World, May 16, 1886.

by any improved or new process of reduction, but was owing simply to the fact that the aluminium industry required sodium, and by making it in large quantities its cost does not exceed the above-mentioned price. The retail price is now $4.00 per pound. The process now used was invented by Brünner, in 1808, and up to the present time nothing new or original has been patented except three or four modifications of his process which have been adopted to meet the requirements of using it on a large scale. Mr. H. Y. Castner, whose laboratory is at 218 West Twentieth Street, New York, has the first patent ever granted on this subject in the United States, and the only one taken out in the world since 1808. Owing to negotiations being carried on, Mr. Castner having filed applications for patents in various foreign countries, but not having the patents granted there yet, we are not at liberty to state his process fully. The metal is reduced and distilled in large iron crucibles, which are raised automatically through apertures in the bottom of the furnace, where they remain until the reduction is completed and the sodium distilled. Then the crucible is lowered, a new one containing a fresh charge is substituted and raised into the furnace, while the one just used is cleaned and made ready for use again. The temperature required is very moderate, the sodium distilling as easy as zinc does when being reduced. Mr. Castner expects to produce it at 25 cents per pound, thus solving the problem of cheap aluminium, and with it magnesium, silicon, and boron, all of which depend on sodium for their manufacture. Thus the production of cheap sodium means much more than cheap aluminium. Mr. Castner is well known in New York as a chemist of good standing, and has associated with him Mr. J. H. Booth and Mr. Henry Booth, both well known as gentlemen of means and integrity."

Mr. Benjamin, in a letter to the 'Engineering and Mining Journal,' gives the following details in addition to those above:* The pots used are cast iron, 8 inches in diameter and 14 inches deep. They are kept at bright red, or about 1000°, at which temperature the decomposition takes place. Whereas, by previous processes only one-third of the sodium in the charge is obtained, Mr. Castner gets nearly all, for the pots are nearly entirely empty when withdrawn from the furnace. Thus the great items of saving are, two or three times as much metal extracted from a given amount of salt, and cheap cast-iron crucibles used instead of expensive wrought-iron retorts.

The following are the claims which Mr. Castner makes in his patent:—†

Claim 1. In a process for manufacturing potassium or sodium, performing the reduction by diffus-

* Eng. and Min. Journ., May 29, 1886.
† U. S. Pat. No. 342,897, June 1, 1886. Hamilton Y. Castner, New York.

ing carbon in a body of alkali in a state of fusion at moderate temperatures.

2. Performing the reduction by means of the carbide of a metal or its equivalent.

3. Mechanically combining a metal and carbon to increase the weight of the reducing material, and then mixing this product with the alkali and fusing the latter, whereby the reducing material is held in suspension throughout the mass of fused alkali.

4. Performing the deoxidation by the carbide of a metal or its equivalent.

We learn later that Mr. Castner cokes a mixture of fine iron and gas-tar, grinds the coke, and uses this as the reducing material; caustic soda is used on account of its low fusing point.

REDUCTION OF SODIUM BY ELECTRICITY.

Mierzinski: In order to lower the cost of sodium efforts have been made to obtain it by means of electricity. Davy has shown that its production in this way is possible, for he first obtained the metal by electrolizing a solution of Na^2CO^3. P. Jablochoff uses the following arrangement to decompose $NaCl$ or KCl:—

The arrangement is easily understood. The salt to be decomposed is fed in by the funnel into the kettle heated by a fire beneath. The positive pole evolves chlorine gas, and the negative pole evolves vapor of the metal, for, as the salt is melted, the heat is sufficient to vaporize the metal liberated. The gas escapes through one tube and the metallic

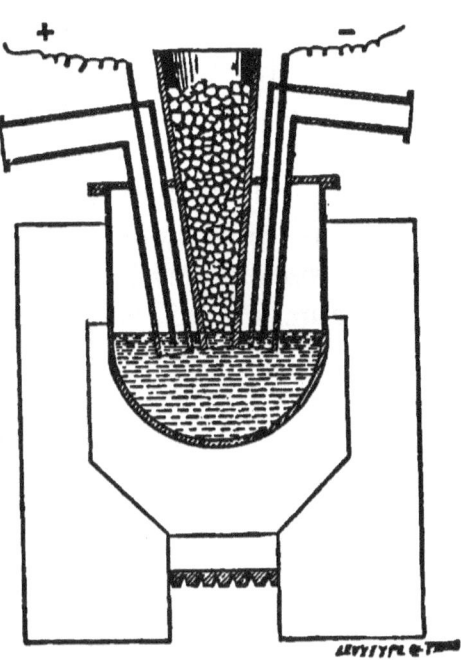

Fig. 5.

vapor by the other. The vapor is led into a condenser and solidified.

PART VII.

MANUFACTURE OF ALUMINA.

I DO not propose to give here all the methods which have been employed to get good clean alumina (Al^2O^3), but only those which may be recommended as being practical and economical on a large scale, not repeating the methods used at Salindres or by Mr. Webster, which will be found in connection with the full description of the processes used at Salindres and Birmingham. Most of the following is from Mierzinski, and may be taken as representing the present state of the industry.

By igniting an alum salt, as ammonia alum, there remains either a white powder or shining, sticky pieces which are very hard and dissolve with difficulty in weak acid or in concentrated solutions of alkali. Large quantities of this alumina may be obtained by calcining the salt in an oven similar in its principal details to a soda furnace.

Mierzinski then gives Mr. Webster's process of mixing the powdered alum with coal-tar, etc., which is given in full in Part IX.

Tilghman decomposes commercial sulphate of alumina, $Al^2(SO^4)^3.18H^2O$, by filling a red-hot fire-clay cylinder with it. This cylinder is lined inside with a magnesia fettling, is kept at a red heat, the sulphate put in in large lumps, and steam is passed through the retort, carrying with it vapor of NaCl. This last arrangement is effected by passing steam into a cast-iron retort in which NaCl is kept melted, and as the steam leaves this retort it carries vapor of the salt with it. It is preferable, however, to make a paste of the sulphate of alumina and the sodium chloride, forming it into small hollow cylinders, which are well dried, and then the fire-clay cylinder filled with these. Then, the cylinder being heated to whiteness, highly superheated steam is passed over it. The HCl which is formed is caught in a condensing apparatus, and there remains a mass of aluminate of soda, which is moistened with water and treated with a current of carbon dioxide and steam. By washing the mass, the soda goes into solution and hydrated alumina remains, which is washed well and is ready for use.

Most of the alumina is now made from the natural aluminous earths, bauxite and cryolite, the occurrence and properties of which have been already described. The manufacture from bauxite is fully described in the account of the process used at Salindres, on p. 158. We will give here the modern methods of making it from cryolite.

Manufacture from Cryolite.

Dry Way.—The cryolite is pulverized, an easy operation, and to every 100 parts, 130 to 150 parts of chalk are added, and a suitable quantity of fluorspar is also used, which remains in the residue on washing after ignition. More chalk is used than is theoretically necessary, in order to make the mass less fusible and keep it porous. But, to avoid using too much chalk merely for this purpose, a certain quantity of coke may be put into the mixture. It is of the first importance that the mixture be very intimate and finely pulverized. It is of greater importance that the mixture be subjected to just the proper well-regulated temperature while being calcined. The cryolite will melt very easily, but this is to be avoided. On this account, the calcination cannot take place in an ordinary smelting furnace, because, in spite of stirring, the mass will melt at one place or another, while at another part of the hearth it is not even decomposed, because the heat at the fire-bridge is so much higher than at the farther end of the hearth. Thomson constructed a furnace for this special purpose (see Figs. 6 and 7), in which the flame from the fire first went under the bed of the furnace, then over the charge spread out on the bed, and finally in a flue over the roof of the hearth. The hearth has an area of nearly 9 square metres, being 4 metres long and 2.5 metres wide. It is charged

MANUFACTURE OF ALUMINA. 147

twelve times each day, each time with 500 kilos of
mixture, thus roasting 6000 kilos daily, with a

Fig. 6.

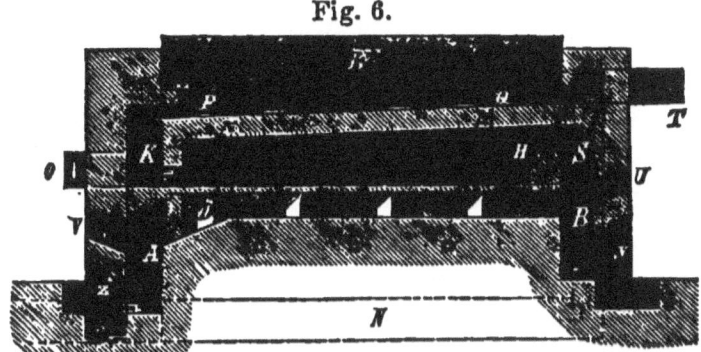

consumption of 800 kilos of coal. The waste heat
of the gases escaping from the furnace is utilized

Fig. 7.

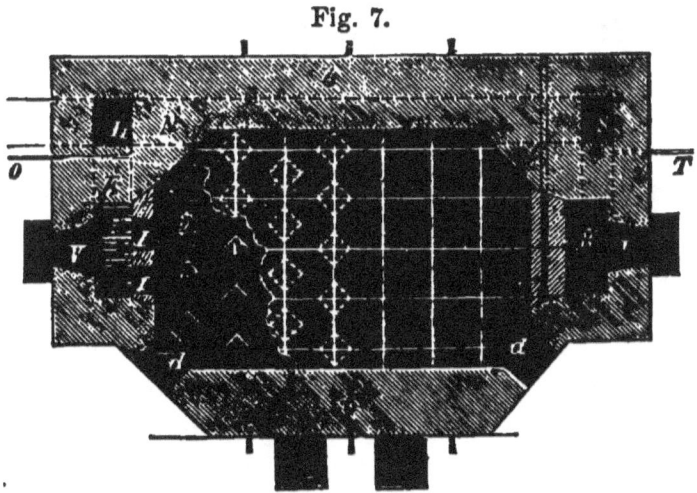

for drying the soda solution to its crystallizing
point, and the gases finally pass under an iron
plate on which the chalk is dried. In this fur-

nace the mass is ignited thoroughly without a bit of it melting, so that the residue can be fully washed with water. The reaction commences at a gentle heat, but is not completed until a red heat is reached. Here is the critical point of the whole process, since a very little raising of the temperature above a red heat causes it to melt. However, it must not be understood that the forming of lumps is altogether to be avoided. These lumps would be very hard and unworkable when cold, but they can be broken up easily while hot, so that they may be drawn out of the furnace a few minutes before the rest of the charge is removed, and broken up while still hot without any trouble. The whole charge, on being taken out, is cooled and sieved, the hard lumps which will not pass the sieve are ground in a mill and again feebly ignited, when they will become porous and may be easily ground up. However, the formation of these lumps can be avoided by industrious stirring of the charge in the furnace. A well-calcined mixture is porous, without dust and without lumps which are too hard to be crushed between the fingers. We would here remark that mechanical furnaces of similar construction to those used in the manufacture of soda, potash, sulphate of soda, etc., are more reliable and give the best results if used for this calcination. The mixture, or ashes, as the workmen call it, is drawn still hot, and washed while warm in conical wooden boxes with double bottoms, or the box may have but one bottom, with an iron plate about 76 millimetres above it. A series of such boxes, or a large apparatus having several compartments, may be so arranged that the washing is done methodically, i. e., the fresh water comes first in contact with a residue which is already washed nearly clean, and the fresh charge is washed by the strong liquor. This is known as the "Lessiveur méthodique," and an apparatus constructed especially for this purpose is described in Dingler 186, 376, by P. J. Havrez, but the subject is too general and the description too long to be given here. A very suitable washing apparatus is also that of Schank, used in the soda industry for washing crude soda, and described in 'Lunge's Handbook of the Soda Industry,' Book II. p. 410. Since the ashes are taken warm from the furnace the washing water need not be previously heated, but the final washwater must be warmed as the ashes have been cooled down by the previous washings. As soon as the strong liquor does not possess a certain strength, say 20° B., it is run over a fresh charge and so brought up. The solution contains sodium aluminate.

Now, whether the sodium aluminate solution is made from beauxite or cryolite, it is treated further in the same way in either case to get the hydrated alumina and the soda solution. Carbon dioxide is next passed through the solution.

The carbon dioxide necessary for precipitating the
13*

hydrated alumina may be made in different ways. The gases coming from the furnace in calcining the cryolite might be used if they were not contaminated with dust; and there is also the difficulty that exhausting the gases from the furnace would interfere with the calcination. It has also been recommended to use the gases from the fires under the evaporating pans, by exhausting the air from the flues and purifying it by washing with water. This can only be done where the pans are fired with wood or gas. However, the lime-kiln is almost exclusively used to furnish this gas. The kiln used is shaped like a small blast furnace. Leading in at the boshes are two flues from five fire-places built in the brickwork of the furnace, and the heat from these calcines the limestone. The gases are taken off by a cast-iron down-take at the top. At the bottom of the furnace, corresponding with the tap hole in a blast furnace, is an opening, kept closed, from which lime is withdrawn at intervals. A strong blast is blown in just above the entrance of the side flues, and by keeping up a pressure in the furnace, leakings into it may be avoided. The gas is sucked away from the top by a pump, which forces it through a cleaning apparatus constructed like a wash bottle, and it is then stored in a gasometer. Instead of the pump, a steam aspirator may be used, which is always cheaper and takes up less room.

The precipitation with CO^2 is made by simply

forcing it through a tube into the liquid. The apparatus used at Salindres is one of the most improved forms. (See p. 163.) The precipitate is granular, and settles easily. However, it is not pure hydrated alumina, but a compound of alumina, soda, carbonic acid, and water, containing usually about 45 per cent. Al^2O^3, 20 per cent. Na^2CO^3, and 35 per cent. H^2O. The sodium carbonate can be separated by long-continued boiling with water, but by this treatment the alumina becomes very gelatinous and very difficult of further treatment. The precipitate was formerly separated on linen filters, but centrifugal machines are now preferred. The evaporated solution gives a high grade of carbonate of soda free from iron. The heavy residue which is left after the ashes have been lixiviated consists of Fe^2O^3, CaO, undecomposed cryolite, and aluminate of Na, and has not been used for anything.

According to Löwig's experiments, the solution of sodium aluminate can be precipitated by calcium, barium, strontium, or magnesium, hydrates, forming caustic soda and hydrated alumina, the latter being precipitated with the CaO, BaO, SrO, or MgO. The precipitate is washed by decantation and then divided into two portions, one of which is dissolved in HCl, the other made into a mush with water and gradually added to the solution of the other half until the filtrate shows only a very little Al^2O^3 in solution. $CaCl^2$, $BaCl^2$, $SrCl^2$, or

$MgCl^2$ has been formed, and the alumina all precipitated.

Wet way.—The decomposition of cryolite in the wet way is operated as follows: The finely powdered cryolite is boiled with dried burnt lime in the proportion of three parts cryolite to two of lime. There results a precipitate of calcium fluoride, CaF^2, while sodium aluminate is in the solution. The reaction takes place quite easily. The solution is settled, washed by decantation, and these washings put with the strong solution first poured off; the next washings are reserved for the fresh wash-water of another operation. The solution of sodium aluminate is then boiled with a quantity of cryolite equal to the amount first used, when sodium fluoride is formed and alumina precipitated. This operation is in no way difficult, only requiring a little more attention than the first. The alumina thus made is very finely divided. The reactions involved are:—

$$Al^2F^6.6NaF + 6CaO = Al^2O^3.3Na^2O + 6CaF^2.$$
$$Al^2F^6.6NaF + Al^2O^3.3Na^2O + 6H^2O = 2(Al^2O^3.3H^2O) + 12NaF.$$

During this last operation it is best to add an excess of cryolite, and keep the liquid in motion to prevent the cryolite from caking at the bottom. Lead is the best material to make these precipitating tanks of, since iron would contaminate the alumina. The precipitate is washed as in the previous operation. The solution of sodium fluoride, NaF, is boiled with the requisite quantity of burnt lime, which converts it into caustic soda, NaOH, which is separated from the precipitated CaF^2 by decantation and washing. The solution is evaporated down to a concentrated solution of NaOH, or to dryness, as desirable. The lime used should be as pure and free from iron as possible, to avoid contaminating the alumina.

Alumina may also be obtained from alum stone or alum shales; by converting these into alums or into $Al^2(SO^4)^3Aq.$ by any of the well-known methods of alum-makers, and then the alumina made by calcining this salt, as given on p. 144.

PART VIII.

MANUFACTURE OF THE DOUBLE CHLORIDE OF ALUMINIUM AND SODIUM.

Since the cost of the aluminium chloride is equal to the cost of the sodium used, in making aluminium, many attempts have been made to cheapen its manufacture from alumina, but without much success. Mr. Frishmuth, of Philadelphia, claims that he has lowered the cost of aluminium principally by making the Al^2Cl^6 much cheaper than before. Mr. Webster's process, which has been applied on such a large scale, is altogether concerned with producing the Al^2Cl^6 cheaply, not a word being said about improvements in making the sodium. However, with these exceptions, and possibly one or two others which will be found further on, the making of these chlorides has remained much as Deville left it. The description of their manufacture as conducted at Salindres is the only account given by Mierzinski, and so that may be taken as the process now in general use, especially in Europe. It will be found on p. 166. I add here just a few words to Fremy's description of the process, which may serve to render his description more exact:—

Mierzinski says that when the double chloride is to be made, special importance is to be attached to using materials free from iron in preparing the alumina, as iron cannot be removed from Al^2Cl^6.-$2NaCl$ as easily as from Al^2Cl^6. Mierzinski also devotes some space to descriptions of chlorine generators, but that is a separate subject, full descriptions of which can be found in any good work on practical chemistry.

There have been a few attempts to make these chlorides in different ways from that used in Deville's process. I find two such processes, which, however, cannot have been much of an improvement on Deville's, or else they would have supplanted it.

M. Dullo makes the following observations on the production of Al^2Cl^6 direct from clay:—*

"Up to the present time the Al^2Cl^6 necessary to the production of aluminium has been prepared by treating cryolite or beauxite, calcining them with carbonate of soda, and neutralizing directly with HCl or CO^2 the aluminate of soda formed. This process may be simplified, and the Al^2Cl^6 obtained much more easily, by direct treatment of clay. For this purpose a good clay, free from iron and sand, is mixed with enough water to make a thick pulp, to which is added NaCl and pulverized carbon. For every 100 parts of dry clay there are

* Bull. de la Soc. Chem. 1860, vol. v. p. 472.

taken 120 parts NaCl and 30 of carbon. The mixture is dried and broken up into small fragments, which are then introduced into a red-hot retort traversed by a current of chlorine. Carbonic oxide is disengaged, while at the same time Al^2Cl^6 and a little $SiCl^4$ are formed. It is not necessary that the chlorine should be absolutely dry, it may be employed just as it comes from the generator. The gas is absorbed very rapidly, because between the aluminium and silicon there are reciprocal actions under the influence of which the chemical actions are more prompt and energetic. The aluminium having for chlorine a greater affinity than silicon has, Al^2Cl^6 is first formed, and it is only when all the aluminium is thus transformed that any $SiCl^4$ is formed. When $SiCl^4$ begins to form the operation is stopped, the incandescent mixture is taken out of the retort and treated with water. The solution is evaporated to dryness to separate out a small quantity of silica, SiO^2, which is in it, the residue is taken up with water, and the $Al^2Cl^6.2NaCl$ remains when the filtered solution is evaporated to dryness. These successive solutions and evaporations might probably be suppressed, especially if only enough chlorine is passed over the incandescent clay to just convert all the aluminium into Al^2Cl^6, in which case no $SiCl^4$ will be formed, and therefore no soluble silica can exist in the solution to contaminate the Al^2Cl^6 or

CHLORIDE OF ALUMINIUM AND SODIUM. 157

impede its reduction. M. Dullo recommends reducing the $Al^2Cl^6.2NaCl$ by zinc. See Part X.

'Chemical News,' 1878, p. 807, contains a short account of an improved method of producing Al^2Cl^6, which consists essentially in passing vapors of hydrochloric acid, HCl, and carbon disulphide, CS^2, simultaneously over heated alumina or clay. The CS^2 changes it into aluminium sulphide, Al^2S^3, and the HCl converts this into Al^2Cl^6, which distils.

PART IX.

Manufacture of Aluminium at Salindres (Gard).

We will now give the actual preparation at Salindres,* with the latest improvements which it has received in practice. Aluminium is there regularly prepared at the works of the Chemical Manufacturing Company of Alais and Camargue, the old firm of Henry Merle & Co., new firm A. R. Pechiney & Co.

The principal chemical reactions on which this process rests are the following:—

Formation of aluminate of soda by calcining beauxite with Na^2CO^3—

$$(AlFe)^2O^3.2H^2O + 3Na^2CO^3 = Al^2O^3.3Na^2O + Fe^2O^3 + 2H^2O + 3CO^2.$$

Formation of alumina by precipitating the aluminate of soda with a current of carbon dioxide—

$$Al^2O^3.3Na^2O + 3CO^2 + 3H^2O = Al^2O^3.3H^2O + 3Na^2CO^3.$$

Formation of $Al^2Cl^6.2NaCl$ by the action of

---

* Fremy's Ency. Chem., M. Margottet.

chlorine on a mixture of alumina, carbon, and sodium chloride—

$Al^2O^3 + 3C + 2NaCl + 6Cl = Al^2Cl^6.2NaCl + 3CO.$

Reduction of this double chloride by sodium.

$Al^2Cl^6.2NaCl + 6Na = 2Al + 8NaCl.$

The primary material then to furnish the aluminium is beauxite. It will be seen that to obtain the metal it is necessary to proceed successively through the following operations:—

I. Preparation of the aluminate of soda and solution of this salt to separate it from the ferric oxide contained in the beauxite.

II. Precipitation of hydrated alumina from the aluminate of soda by a current of carbon dioxide; washing the precipitate.

III. Preparation of a mixture of alumina, carbon, and salt, drying it, and then treating with gaseous chlorine to obtain the double chloride of aluminium and sodium.

IV. Lastly, treatment of this chloride by sodium to obtain aluminium.

We will now review these operations as practically carried out in detail. We will not consider the preparation of the crude materials as chlorine, sodium, etc., which is spoken of elsewhere.

I. *Preparation of the Aluminate of Soda.*

The aluminate to serve for the preparation of $Al^2Cl^6.2NaCl$ was first obtained by the calcination of ammonia alum. At Salindres this was withdrawn and beauxite used, a material consisting of sesqui-oxide of iron and aluminium in varying proportions, with two molecules of water and a little silica. It is redder the more iron it contains. Beauxite is plentiful enough in the south of France, principally in the departments of Herault, Bouches-du-Rhone, and Var. That used at Salindres comes from Var. It contains at least seventy-five per cent. alumina. To separate the alumina from Fe^2O^3, it is treated with carbonate of soda, under the influence of a sufficiently high temperature, the Al^2O^3 displacing the CO^2 and forming aluminate of soda, $Al^2O^3.3Na^2O$, while the Fe^2O^3 remains unattacked. A simple washing with water then permits the separation of the $Al^2O^3.3Na^2O$ from the insoluble Fe^2O^3. The beauxite is first finely pulverized by means of a vertical mill-stone, then intimately mixed with some Na^2CO^3. The mixture is made for one operation, of—

    480 kilos beauxite.
    300 " Na^2CO^3 of 90 alkali degrees.

This mixture is introduced into a reverberatory furnace, resembling in form a soda furnace, and which will bear heating strongly. The mass is stirred from time to time, and it is kept heated until all the carbonate has been attacked, which is recognized by a test being taken which does not effervesce with acids. The operation lasts from five to six hours.

REDUCTION OF THE ALUMINIUM. 161

The aluminate thus obtained is separated from Fe^2O^3 by a washing with warm water. This washing is made at first with a feeble solution which has served for the complete exhaustion of the preceding charge, which was last washed with pure water, forming thus this feeble solution. This gives, on the first leaching, solutions of aluminate concentrated enough to be called strong liquor, which are next treated by the current of CO^2 to precipitate the hydrated alumina. The charge is next washed with pure water, which completely removes the aluminate; this solution is the weak liquor, which is put aside in a special tank, and used as the first leaching liquor on the next charge treated. This treatment takes place in the following apparatus (see Fig. 8): B is a sheet-iron vessel, in the middle of which is a metallic grating, F, on which is held all round its edges, by pins, a cloth, serving as a filter. The upper part of this vessel is called simply the filter. A ought to be closed by a metallic lid held on firmly by bolts. To work the apparatus, about 500 kilos of the charge to be washed is placed on the filter cloth, the lid is closed, then the steam-cock f of the reservoir A is opened. In A is the weak solution from the last washing of the preceding charge. The pressure of the steam makes it rise by the tube T into the filter; another jet of steam, admitted by the cock b, rapidly warms the feeble liquor as it soaks into the charge. After filtering through, the strong liquor is drawn off by turning the stopcock G. The weak solution of the reservoir A is put into the filter in

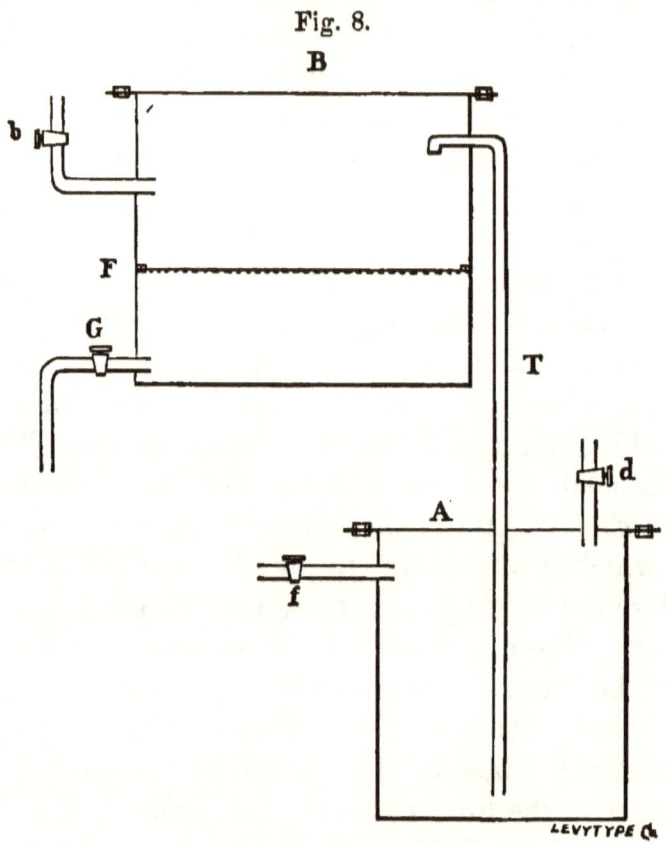

Fig. 8.

successive portions, and not all at once; and after each addition of solution has filtered through, its strength in B.° is taken, before any more solution is run in; then, when the solution marks 3 to 4°, it is placed in the special tank for weak liquor, with all that comes through afterwards. Just about this time, the weak liquor of the reservoir A is generally all used up, and is replaced by pure water introduced by the tube d. All the solutions which filtered through, marking over 3 to 4° B., are put together, and form the strong liquor which marks about 12° B. This extraction of the aluminate being completed by the pure water, the residue on the filter is taken out, and a new operation may be commenced.

II. *Preparation of the Alumina.*

The strong liquor is introduced into a vessel having an agitator, where a strong current of CO^2 may precipitate the Al^2O^3 from it. The gas is produced by small streams of hydrochloric acid continuously falling on some limestone contained in a series of earthenware jars. The precipitation vessel is called a baratte. The CO^2 after having passed through a washing flask, is directed to a battery of three barattes, where the precipitation is worked methodically, so as to precipitate completely the alumina of each baratte, and utilize at the same time all the carbon dioxide produced. In order to do this, the gas always enters first into a baratte in which the precipitation is nearest completion, and arrives at last to that in which the solution is freshest. When the gas is not all absorbed in the last baratte, the first is emptied, for the precipitation in it is then completed, and it is made the last of the series, the current being now directed first into the baratte which was previously second, while the newly charged one is made the last of the series. The process is thus kept on

Fig. 9.

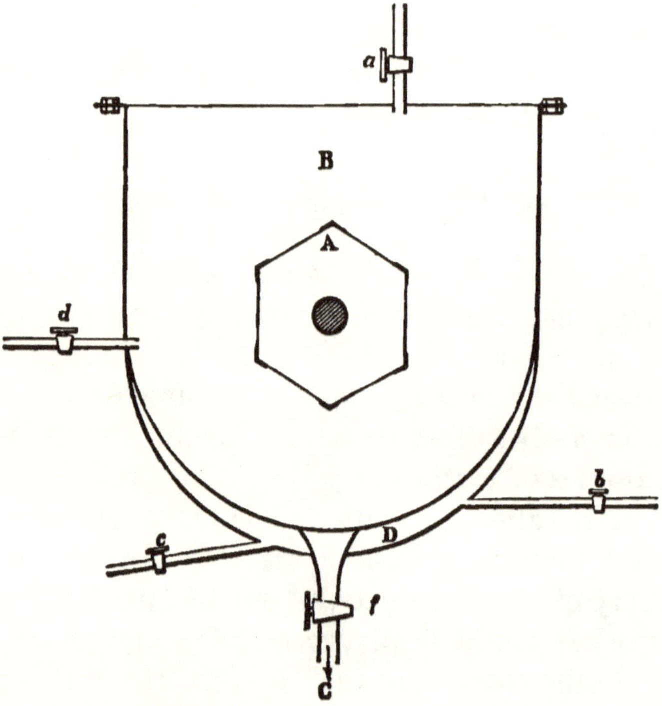

a. Charging pipe.
b. Steam pipe.
c. Steam drip.
d. CO^2 enters.
f. Discharge pipe.
A. Agitator, made of iron rods.
C. Tank in which the precipitate settles.
B. Baratte body.
D. Steam jacket.

continuously. The apparatus used is shown in Fig. 9.

Each baratte holds about 1200 litres of solution, and the complete precipitation of all the alumina in it takes fivê to six hours. A mechanical agitator stirs the contents continually, and a current of steam is let into the double bottom so as to keep the temperature of the solution about 70°. The precipitated alumina and the solution of Na^2CO^3 which remains are received in a vat placed beneath each baratte. The solution is decanted off clear, after standing, and then evaporated down to dryness, regenerating the Na^2CO^3 used in treating the beauxite to make the aluminate, less the inevitable losses inseparable from all industrial operations. The deposit of alumina is put into a conical strainer to drain, or else into a centrifugal drying machine, which rapidly drives out of the hydrated alumina the solution of Na^2CO^3 which impregnates it; a washing with pure water in the drier itself terminates the preparation of the alumina. At the works at Salindres, a part of this alumina is converted into sulphate of alumina, which is sold, the remainder being used for the aluminium manufacture. After washing in the dryer, the alumina presents this composition:—

Al^2O^3	47.5
H^2O	50.0
Na^2CO^3	2.5

III. *Preparation of the $Al^2Cl^6.2NaCl$.*

When a current of chlorine is passed through a mixture of anhydrous alumina and carbon, Al^2Cl^6 is obtained. This simple chloride may be employed for obtaining aluminium; it was first so employed by Deville; but it is deliquescent, its preservation is difficult, and its employment very inconvenient. Industrially, as indicated by Deville, the double chloride is always used, as it does not present these inconveniences to so large a degree. The double chloride may be obtained in the same manner as the simple chloride; it is sufficient to put some common salt, NaCl, into a mixture of alumina and carbon, and, on heating this mixture strongly, there is formed, by the action of the chlorine, $Al^2Cl^6.2NaCl$, which distils at a red heat and condenses in a crystalline mass at about 200°. The hydrated alumina obtained in the preceding operation is mixed with salt and finely pulverized charcoal, in proper proportions, the whole is sifted, and a mixture produced as homogeneous as possible; then it is agglomerated with water and made into balls the size of the fist. These balls are first dried in a drying stove, at about 150°, then calcined at redness in retorts, where the double chloride should commence to be produced just as the balls are completely dried. These retorts are vertical cylinders of refractory earth, each one is furnished with a tube in its lower

REDUCTION OF THE ALUMINIUM.

part for the introduction of chlorine, and with another towards its upper end for the exit of the vapor of double chloride. (See Fig. 10.) A lid

Fig. 10.

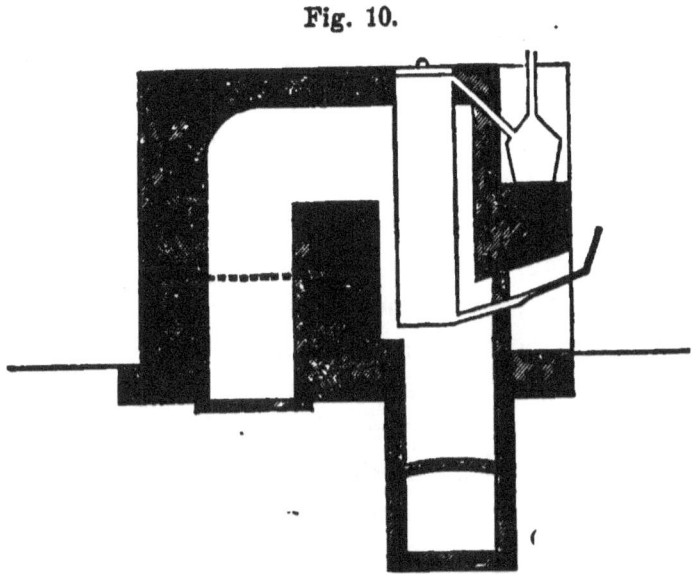

carefully luted during the operation with a mixture of fine clay and horse dung serves for the charging and discharging of the retort. The double chloride is condensed in earthen pots like flower pots, made of ordinary clay, and closed by a well-luted cover, into which passes a pipe of clay to conduct the gas resulting from the operation into flues connected with the main chimney. Each retort is heated by a fire, the flame of which circulates all round it, and permits keeping it at a bright red heat. An operation is conducted as follows:

The retort is filled with stove-dried balls, the lid is carefully luted, and the retort is heated gently till all the moisture is driven off. This complete desiccation is of great importance, and requires much time. Then chlorine, furnished by a battery of three generating vessels, is passed in. During the first hours, the gas is totally absorbed by the balls, and the double chloride distils regularly for about three hours, and runs into the earthen pots where it solidifies. Toward the end, the distillation is more difficult and less regular, and the chlorine is then only incompletely absorbed. After each operation there remains a little residue in the retort, which accumulates and is removed every two days, when two operations are made per day. One operation lasts at least twelve hours, and a retort lasts sometimes a month. The double chloride is kept in the pots in which it was condensed until the time it is to be used in the next operation; it is almost chemically pure, save traces of iron, and is easy to keep and handle.

IV. *Reduction of the Double Chloride by Sodium.*

The difficulty of this operation, at least from an industrial point of view, is to get a slag fusible enough and light enough to let the reduced metal easily sink through it and unite. This result has been reached by using cryolite, a white or grayish mineral originally from Greenland, very easy to melt, formula $Al^2F^6.6NaF$. This material forms with the NaCl resulting from the reaction a very fusible slag, in the midst of which the aluminium collects well, and falls to the bottom. In one operation the charge is—

100 kilos	$Al^2Cl^6.2NaCl$.
45 "	$Al^2F^6.6NaF$.
35 "	Na.

The double chloride and cryolite are pulverized, the sodium, cut into small pieces a little larger than the thumb, is divided into three equal parts, each part being put into a sheet-iron basket. The mixture of double chloride and cryolite, being pulverized, is divided into four equal parts, three of these are respectively put in each basket with the sodium,

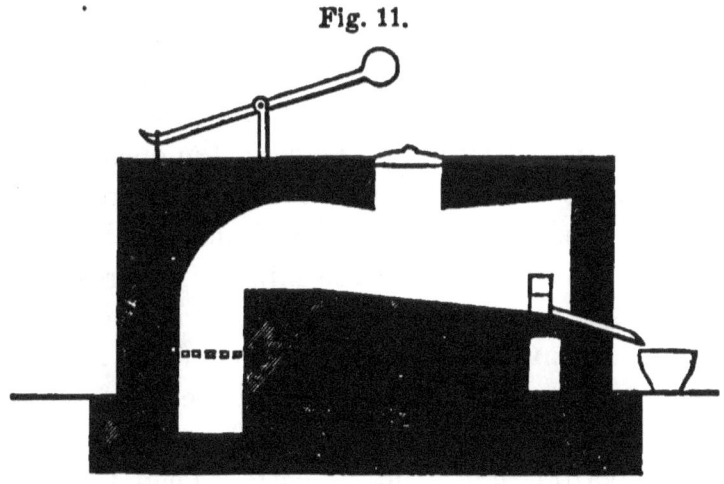

Fig. 11.

the fourth being placed in a basket by itself. The reduction furnace (see Fig. 11) is a little furnace of refractory brick, with an inclined hearth and a vaulted roof. This furnace is strongly braced by iron tie-rods, because of the concussions caused by the reaction. The flame may at any given moment be directed into a flue outside of the hearth. At the back part of the furnace, that is to say, on that side towards which the bed slopes, is a little brick wall which is built up for each reduction and is taken away in operating the running out of the metal and slag. A gutter of cast iron is placed immediately in front of the wall to facilitate running out the materials. All this side of the furnace ought to be opened or closed at pleasure by means of a damper. Lastly, there is an opening for charging in the roof, closed by a lid. At the time of an operation the furnace should be heated to low redness, then are introduced in rapid succession the contents of the three baskets containing sodium, etc., and lastly the fourth containing only double chloride and no sodium. Then all the openings of the furnace are closed, and a very vivid reaction accompanied by dull concussions immediately takes place. At the end of fifteen minutes, the reaction subsides, the dampers are opened, and the heat continued, meanwhile stirring the mass from time to time with an iron poker. At the end of three hours the reduction is ended, and the metal collects at the bottom of the liquid bath. Then the running out is proceeded with in three phases: First. Running off the upper part of the bath, which

REDUCTION OF THE ALUMINIUM. 171

consists of a fluid material completely free from reduced aluminium and constituting the white slag. To run this out a brick is taken away from the upper course of the little wall which terminates the hearth. These slags are received in an iron wagon. Second. Running out the aluminium. This is done by opening a small orifice left in the bottom of the brick wall, which was temporarily plugged up. The liquid metal is received in a cast-iron melting pot, the bottom of which has been previously heated to redness. This aluminium is immediately cast in a series of small rectangular cast-iron moulds. Third. Running out of the rest of the bath, which constitutes the gray slags. These were, like the white slags, formed by the NaCl and cryolite, but they contain in addition, isolated globules of aluminium. To run these out all the bricks of the little wall are taken away. This slag is received in the same melting-pot into which the aluminium was run, the latter having been already moulded Here it cools gradually, and after cooling there are always found at the bottom of the pot several grains of metal. In a good operation there are taken from one casting 10.5 kilos of aluminium, which is sold directly as commercial metal.

The foregoing description from Fremy sets forth in its perfection the production of aluminium by means of sodium, and until very recently this was the only successful commercial process. A large amount of aluminium is now produced by this process, and it, therefore, does not lack interest. The following data as to the expense of this process may be very appropriately inserted here, giving the cost at Salindres in 1872.

In 1872, 3600 kilos of Al were made at Salindres at the following average cost:—

α. Manufacture of one kilo of Na.

Soda	9.35 kilos @ 32 fr. per 100 kilos	= 3 fr. 9 cent.
Coal	74.32 " " 1.40 " " " "	= 1 " 4 "
	Wages . . .	1 " 73 "
	Expenses . .	3 " 46 "
	Total	11 " 32 "

β. Manufacture of one kilo of $Al^2Cl^6.2NaCl$.

Anhydrous Al^2O^3

	0.59 kilos @ 86 fr. per 100 kilos	= 0 fr. 50.7 cent.
MnO^2	3.74 " " 14 " " " "	= 0 " 52.3 "
HCl	15.72 " " 3 " " " "	= 0 " 47.1 "
Coal	25.78 " " 1.40 " " " "	= 0 " 36.1 "
	Wages . . .	0 " 23.8 "
	Expenses . .	0 " 38.0 "
	Total	2 " 48.0 "

γ. Manufacture of one kilo of Al.

Na . . 3.44 kilos @ 11.32 fr. per kilo	= 38 fr. 90 cent.
$Al^2Cl^6.2NaCl$ 10.04 " " 2.48 " " "	= 24 " 90 "
Cryolite 3 87 " " 61.0 " " 100 kilos	= 2 " 36 "
Coal 29.17 " " 1.40 " " " "	= 0 " 41 "
Wages . . .	1 " 80 "
Costs . . .	0 " 88 "
Total . . .	69 " 25 "

---

* A. Wurtz, Wagner's Jaresb., 1874, vol. xxi.

This must be increased ten per cent. for losses and other expenses, making the cost of aluminium 80 fr. per kilo, and it is sold for 100.

According to a statement in the 'Bull. de la Soc. de l'Industrie Minérale,' ii., 451, made in 1882. Salindres was then the only place in which aluminium was manufactured.

Later Improvements in Deville's Processes.

The later improvements in this process have been made principally by Mr. J. Webster, of Birmingham, England, and some are claimed by Frishmuth of Philadelphia. We will examine the reports of Webster's processes and the claims of Frishmuth.

Webster's Process.

Recently the statement* has been current in a number of journals that material improvements have been made in the manufacture of aluminium at the Aluminium Crown Metal Works at Hollywood, near Birmingham, England, under the direction of Mr. Webster. Mr. Webster describes one of his improvements, which is patented,† as follows: Three parts of alum are mixed with one part of coal pitch, and the mixture heated to 200° or 260°.

---

* Dingler, 1883, cclix. 86.
† Austrian Pat. Sept. 28, 1882.

In about three hours the pasty mass is spread upon a stone floor, and after becoming cool is broken in pieces. Hydrochloric acid of twenty to twenty-five per cent. is poured upon these pieces placed in piles which are turned over from time to time. When the evolution of sulphuretted hydrogen, H^2S, has stopped, about five per cent. of charcoal powder or lampblack, with enough water to make a thick paste, is added. The mass is thoroughly broken up and mixed in a mill, and then worked into balls of about a pound each. These are bored through to facilitate drying, and heated in a drying chamber at first to 40°, then in a furnace from 95 up to 150°. The balls are then kept for three hours at a low red heat in retorts while a mixture of two parts steam and one part air is passed through, so that the sulphur and carbon are converted into SO^2 and CO^2, and thus escape. The current of gas carries over some K^2SO^4, $FeSO^4$, and Al^2O^3, and is therefore passed through clay condensers. After these have been driven off the dry residue is removed from the retort, again ground in a mill to fine powder, which now consists of Al^2O^3 and K^2SO^4. This powder is treated with about seven times its weight of water, then boiled in a pan or boiler by means of steam for about one hour, then allowed to stand till cool. The solution containing the K^2SO^4 is run off and evaporated to dryness, the alumina is washed out and dried.

REDUCTION OF THE ALUMINIUM. 175

The product thus obtained contains 84.1 per cent. Al^2O^3.

The above patent is seen to cover only the manufacture of pure alumina. A later account thus describes Mr. Webster's plant and processes. It is taken from the Birmingham, England, 'Gazette,' and was copied into an American journal* as follows:—

"There has been recently patented in most of the leading countries of the world an invention of great importance. The Aluminium Crown Metal Co., at Hollywood, near Birmingham, now claim to have perfected an improved process by which they produce pure alumina from alum, convert it into Al^2Cl^6, and reduce this by sodium. By this process the two common impurities of aluminium, silicon and iron, are avoided. The inventor is Mr. James Webster, the founder and principal of the company. Their works having been erected within the last five years, the plant is of the most recent date, comprising all the modern improvements in calcining furnaces and retorts, sheet-rolling and wire-drawing mills, together with the requisite casting, fitting, and other shops.

"On retiring from business some years ago as a metal manufacturer, Mr. Webster took up his residence at Hollywood, and while nominally engaged

---

* Bulletin of the Iron and Steel Association, Philadelphia, January 3, 1883.

in farming carried on the experiments which he had commenced as far back as 1851 for the invention of an expeditious and inexpensive mode of producing aluminium. He designed all the various buildings, appliances, and apparatus necessary for the carrying on of experiments, upon which he expended upwards of £3000, besides £2000 or £3000 in procuring patent rights at home and abroad. A French syndicate has just offered him £25,000 for the patent for France alone, while parties in the United States, Belgium, and Germany are arranging to purchase rights.

"The invention has only been perfected about eighteen months, and the firm have but recently begun to place the product on the market, yet such is the demand, that though they are now working day and night, they cannot execute one-quarter of the orders accumulating on their books. By the ordinary method of precipitation, 12 tons of alum and 6 tons of K^2CO^3 or Na^2CO^3 are required to produce one ton of alumina, and the whole process occupies nine weeks; whereas, in Mr. Webster's plan, no precipitant is used, and a ton can be manufactured in a week with the existing plant. The cost of one ton of alumina by the ordinary method is upwards of £1000, while it is less than £100 by Mr. Webster's process.

"Mr. Webster's process consists in taking a given quantity of alum and pitch, which are finely ground, mixed together, and placed in a calcining furnace, by which means 38 per cent. of water is driven off, leaving the sulphur, potash, and alumina, with some ferric oxide. The calcined mixture is then put in vertical retorts, and steam and air are forced through, which leaves a residue of K^2O and Al^2O^3 only. This is then placed in a vat of warm water heated by steam. The caustic potash liquor is then run off and boiled down, while the residual Al^2O^3 is collected in sacks and dried. This deposit contains about 84 per cent. Al^2O^3, while that obtained by the old process of precipitation has only 65 per cent. Thus a saving is effected of nine-tenths in cost and 19 per cent. more alumina is obtained. In addition to this, the whole of the bye products are recovered, consisting of KOH,S (which is used in making H^2SO^4), and aluminate of iron. From these bye products is made a blue dye, which is sold for six shillings a pound, and is used in place of indigo for dyeing calico and other materials. The Al^2Cl^6 is reduced by sodium.

"'The English Ironmonger' for April, 1886, contains a long article describing the extensions which this company have made, their works having now attained a large size, while the number and variety of their products, in aluminium and its alloys, as ingots, wire, sheet, or worked up in hundreds of different ways, is truly surprising. They have monopolized this business in England, and are very enterprising in introducing their manufactures elsewhere."

Frishmuth's Process.

In the United States the only improvement in the sodium process of reducing aluminium is contained in the following patent:—*

Wm. Frishmuth, of Philadelphia, in his patent makes the following claims :—

1. The simultaneous generation of sodium vapor and a volatile compound of aluminium in two separate vessels or retorts, and mingling the vapors thus obtained in a nascent (?) state in a third vessel, wherein they react on each other.

2. The sodium vapor is produced from a mixture of a sodium compound and carbon, or some other reducing agent; and the aluminous vapor from aluminous material.

3. The simultaneous generation of sodium vapor and vapor of Al^2Cl^6 or Al^2F^6; or of sodium vapor and $Al^2Cl^6.2NaCl$.

4. Converting the aluminous material to a vapor by heating it in a retort with NaCl, and subjecting it at the same time to chlorine gas; mingling the vapor of $Al^2Cl^6.2NaCl$ thus obtained with vapor simultaneously generated from Na^2CO^3 and carbon.

* U. S. Pat., 308,152. Nov. 18, 1884.

Other Processes.

II. Niewerth, of Hanover, has patented in the United States and other countries the following process :* A compound of aluminium, with chlorine or fluorine, is brought by any means into the form of vapor, and conducted, strongly heated, into contact with a mixture of 62 parts Na^2CO^3, 28 coal and 10 chalk, which is also in a highly heated condition. This mixture disengages sodium, which reduces the gaseous chloride or fluoride of aluminium, the nascent sodium being the reducing agent. In place of the above mixture other suitable mixtures which generate sodium may be employed, or mixtures may also advantageously be used from which potassium is generated.

Hector von Grousilliers, Springe, Hanover, patents the following improvement :† In order to avoid the difficulties ordinarily met with in the use of $Al^2Cl^6.2NaCl$ to obtain aluminium, the patentee raises the volatilizing point of Al^2Cl^6 by performing its reduction, either chemically or electrolytically, under pressure in a strong, hermetically-closed vessel lined with clay or magnesia and provided with a safety valve.

* Sci. Am. Suppl., Nov. 17, 1883.
† Eng. Pat., June 29, 1885, No. 7858.

PART X.

REDUCTION OF ALUMINIUM BY OTHER REDUCING AGENTS THAN SODIUM.

Reduction by Cyanogen.

According to Knowles's patent,* aluminium chloride, Al^2Cl^6, is reduced by means of potassium or sodium cyanide, the Al^2Cl^6, either fused or in the form of vapor, being brought in contact with either the melted cyanide or its vapor. The patent further states the strange fact that pure alumina may be added to increase the product.

Corbelli, of Florence,† patented the following method in England: Common clay is freed from all foreign particles by washing, then well dried. One hundred grammes of it are mixed with six times its weight of concentrated sulphuric or hydrochloric acid; then the mixture is put in a crucible and heated to 400 or 500°. The mass resulting is mixed with 200 grammes of dry yellow prussiate of potash and 150 grammes of NaCl, and this mixture heated in a crucible to whiteness.

* Sir Francis C. Knowles, Eng. Pat. 1857, No. 1742.
† Wagner's Jahresb., 1858.

After cooling, the reduced aluminium is found in the bottom of the crucible as a button.

According to Deville's experiments, this process will not give any results. Watts remarks that any metal thus obtained must be very impure, consisting chiefly of iron. The patent is dated 1858, No. 142.

Reduction by Hydrogen.

F. W. Gerhard* decomposes aluminium fluoride, Al^2F^6, or $Al^2F^6.6NaF$—cryolite—by subjecting it to hydrogen at a red heat. The aluminium compound is placed in a number of shallow dishes of glazed earthenware, each of which is surrounded by a number of other dishes containing iron filings. These dishes are placed in an oven previously heated to redness, hydrogen gas is then admitted, and the heat increased. Aluminium then separates, hydrofluoric acid, HF, being formed, but immediately taken up by the iron filings and thereby prevented from reacting on the aluminium. To prevent the pressure of the gas from becoming too great, an exit tube is provided, which may be opened or closed at pleasure. This process, patented in England in 1856, No. 2980, is ingenious and was said to yield good results. The inventor has, however, returned to the use of the

* Watts's Dictionary.

more costly reducing agent, sodium, which would seem to imply that the hydrogen method has not yet quite fulfilled his expectations.

Reduction by Carburetted Hydrogen.

Mr. A. L. Fleury,[*] of Boston, mixes pure alumina with gas tar, resin, petroleum, or some such substance, making it into a stiff paste which may be divided into pellets and dried in an oven. They are then placed in a strong retort or tube which is lined with a coating of plumbago. In this they are exposed to a cherry-red heat. The retort must be sufficiently strong to stand a pressure of from 25 to 30 pounds per square inch, and be so arranged that by means of a safety valve the necessary amount of some hydro-carbon may be introduced into the retort among the heated mixture, and a pressure of 20 to 30 pounds must be maintained. The gas is forced in by a force pump. By this process the Al^2O^3 is reduced, while the metal remains as a spongy mass mixed with carbon. This mixture is re-melted with metallic zinc, and when the latter has collected the aluminium, it is driven off by heat. The hydrocarbon gas under pressure is the reducing agent. The time required for reducing 100 pounds of alumina, earth, cryolite, or other compound of aluminium,

[*] Chemical News, June, 1869, p. 332.

should not be more than four hours. When the gas can be applied in a previously heated condition as well as being strongly compressed, the reduction takes place in a still shorter period.

Nothing is now heard of this process, and it has been presumably a failure. It is said that several thousand dollars were expended by Mr. Fleury and his associates without making a practical success of it. We should be glad to hear in the future that their sacrifices have not been in vain, and that the process still has possibilities in it which will some time be realized.

Petitjean* makes aluminium sulphide, Al^2S^3, by one of Fremy's methods,† or makes a double sulphide of aluminium with potassium or sodium by mixing alumina with a little tar or turpentine in a carbon lined crucible, heating strongly, and then mixing with a powder composed of Na^2CO^3, or K^2CO^3, and sulphur; again heating a long time at bright redness. The sulphide or double sulphide thus made is put in a crucible or retort through the bottom of which can be led a stream of carburetted hydrogen, which separates the aluminium from its combination with the sulphur. Aluminium‡ can also be reduced from Al^2S^3 by mixing it with iron filings or a pulverized metal having

* Kerl and Stohman, Poly. Central Blatt. 1858, 888.
† See Appendix.
‡ See Appendix.

similar qualities, and melting the mixture. A metallic mixture may be used instead of carburetted hydrogen in the above operation.

Reduction by Double Reaction.

M. Comenge,* of Paris, obtains aluminium from its sulphide either by heating it in an atmosphere of hydrogen, or by heating it with Al^2O^3 or $Al^2(SO^4)^3$ in such proportions that sulphur dioxide, SO^2, and aluminium may be the sole products; or the sulphide may be decomposed by iron, copper, or zinc. The reactions involved would be—

$Al^2S^3 + 3H-H = 2Al + 3H^2S$.
$Al^2S^3 + 2Al^2O^3 = 6Al + 3SO^2$.
$Al^2S^3 + Al^2(SO^4)^3 = 4Al + 6SO^2$.
$Al^2S^3 + 3(Fe.Cu.Zn.) = 2Al + 3(Fe.Cu.Zn.)S$.

Johnson† patented the following process: Aluminium sulphide is mixed with quite dry $Al^2(SO^4)^3$ in such proportions that the sulphur and oxygen present may evolve as SO^2. The mixture is heated to redness in an unoxidizing atmosphere, when SO^2 evolves and the metal remains. The reaction is furthered by agitation. The aluminium in the resulting mass can be treated in the way commonly used in puddling spongy iron, and then either pressed or hammered together. Or, the aluminium

* Eng. Pat. 1858, No. 461.
† Kerl and Stohman's Handbuch.

sulphide may be heated to redness in an unoxidizing atmosphere and dry hydrogen or water gas conducted over it, and the metal separated from the resulting mass by dressing.

Mr. Niewerth's* process may be operated in his newly invented furnace, but it may also be carried on in a crucible or another form of furnace. The furnace alluded to consists of three shaft furnaces, the outer ones well closed on top by iron covers, and connected beneath by tubes with the bottom of the middle one: the tubes being provided with closing valves. These side shafts are simply water-gas furnaces, delivering hot water-gas to the central shaft, and by working the two alternately supplying it with a continuous blast. The two producers are first blown very hot by running a blast of air through them with their tops open, then the cover of one is closed, the blast shut off, steam turned on just under the cover, and water gas immediately passes from the tube at the bottom of the furnace into the central shaft. The middle shaft has meanwhile been filled with these three mixtures in their proper order:—

First. A mixture of sodium carbonate, carbon, sulphur, and alumina.

Second. Aluminium sulphate.

Third. A flux, preferably a mixture of NaCl and KCl.

* Sci. Am. Suppl., Nov. 17, 1885.

This central shaft must be already strongly heated to commence the operation, it is best to fill it with coke before charging, and as soon as that is hot to put the charges in on the coke. Coke may also be mixed with the charges, but it is not necessary. The process then continues as follows: The water-gas enters the bottom of the shaft at a very high temperature. These highly heated gases, carbonic oxide and hydrogen, act upon the charges so that the first breaks up into a combination of sodium sulphide and aluminium sulphide, from which, by means of the second charge of $Al^2(SO^4)^3$, free aluminium is reduced. As the latter passes down the shaft, it is melted and the flux assists in collecting it, but is not absolutely necessary. Instead of producing this double sulphide, pure aluminium sulphide might be used for the first charge, or a mixture which would generate Al^2S^3; or, again, pure Na^2S, K^2S, CuS, or any other metallic sulphide which will produce the effect alone, in which case aluminium is obtained alloyed with the metal of the sulphide. Instead of the first charge a mixture of alumina, sulphur, and carbon might be introduced. Or, the $Al^2(SO^4)^3$ of the second charge might be replaced by alumina. So, one charge may be Na^2S, K^2S, or any other metallic sulphide, and the second charge may be either Al^2O^3 or $Al^2(SO^4)^3$.

Reduction by Carbon and Carbon Dioxide.

J. Morris* of Uddington claims to obtain aluminium by treating an intimate mixture of alumina and charcoal with carbon dioxide. For this purpose, a solution of Al^2Cl^6 is mixed with powdered wood-charcoal or lampblack, then evaporated till it forms a viscous mass which is shaped into balls. During the evaporation hydrochloric acid is given off. The residue consists of alumina intimately mixed with carbon. The balls are dried, then treated with steam in appropriate vessels for the purpose of driving off all the chlorine, care being taken to keep the temperature so high that the steam is not condensed. The temperature is then raised so that the tubes are at a low red heat, and dry carbon dioxide, CO^2, is then passed through. This CO^2 is said to be reduced by the carbon to carbonic oxide, CO, which now, as affirmed by Mr. Morris, reduces the alumina. Although the quantity of carbonic oxide escaping is in general a good indication of the progress of the reduction, it is, nevertheless, not advisable to continue heating the tubes or vessels until the evolution of this gas has ceased, as in consequence of slight differences in the consistency of the balls some of them give up all their carbon sooner than others. The treatment with carbon

* Dingler, 1883, vol. 259, p. 86. German Pat. No. 22150, Aug. 30, 1882.

dioxide lasts about thirty hours when the substances are mixed in the proportion of five parts carbon to four parts alumina. Morris states further that the metal appears as a porous spongy mass, and is freed from the residual alumina and particles of charcoal either by smelting it, technically "burning it out," with cryolite as a flux or by mechanical treatment.

REDUCTION BY CARBON.

About the first attempt of this nature we can find record of is the following article by M. Chapelle :—*

"When I heard of the experiments of Deville, I desired to repeat them, but having neither aluminium chloride nor sodium to use, I operated as follows: I put natural clay, pulverized and mixed with ground NaCl and charcoal, into an ordinary earthen crucible and heated it in a reverberatory furnace, with coke for fuel. I was not able to get a white heat. After cooling, the crucible was broken, and gave a dry pulverulent scoria in which were disseminated a considerable quantity of small globules about one-half a millimetre in diameter, and as white as silver. They were malleable, insoluble in nitric or cold hydrochloric acids, but at 60° dissolved rapidly in the latter with evolution of hydrogen; the solution was colorless and gave

* Compt. Rendus, 1854, vol. xxxviii. p. 358.

with ammonia a gelatinous precipitate of hydrated alumina. My numerous occupations do not permit me to assure myself of the purity of the metal. Moreover, the experiment was made under conditions which leave much to be desired, but my intention is to continue my experiments and especially to operate at a higher temperature. In addressing this note to the Academy I but desire to call the attention of chemists to a process which is very simple and susceptible of being improved. I hope before many days to be able to exhibit larger globules than those which my first experiment furnished."

M. Chapelle never did address any further communications to the Academy on this subject, and we must presume that further experiments did not confirm these first ones.

G. W. Reinar* states that the pyrophorous mass which results from igniting potash or soda alum with carbon, contains a carboniferous alloy of aluminium with potassium or sodium, from which the alkaline metal can be removed by weak nitric acid.

Cowles Bros.' Process.

This process, which reduces alumina by carbon in the presence of another metal to take up the aluminium, using the electric furnace, is the nov-

* Wagner's Jahresb. 1859, p. 4.

elty which is attracting widespread attention to the metallurgy of aluminium. Its history has already been sketched, and will be still further developed in the following pages. It properly comes under the heading of "Reduction by Carbon."

"Early in the present century, Sir H. Davy, Berzelius, and Oerstedt, all famous chemists, attempted unsuccessfully to reduce alumina by electricity. Likewise, many learned scientists have striven to decompose it by carbon, as other metals are smelted from their ores, but without success, and the opinion has become profound and widespread among chemists that alumina could not be reduced by carbon and heat. But this is exactly what the Cowles process accomplishes, and by its means the Cowles Electric Smelting and Aluminium Company is enabled to supply the alloys of aluminium with other metals at one-quarter to one-third their former price. As to the details of the process, we refer to the papers of Professor Hunt and the one read by Mr. Mabery."*

The following is the patent claim of Messrs. Cowles: U. S. Pat. 324,658 and 324,659, August 18, 1885. Electric smelting of aluminium. To Cowles Bros., Cleveland, Ohio. Claim: Reducing the aluminium compound in company with a metal in a furnace heated by electricity in presence of

* Cowles Bros.' Pamphlet.

carbon. The alloy of aluminium and the metal formed is treated to separate the aluminium.

The following paper is the first official and scientific account of Cowles Bros.' process, and was read before the American Association for the Advancement of Science by Professor Charles F. Mabery of the Case School of Applied Science, Cleveland.*

" The application of electricity to metallurgical processes has hitherto been confined to the reduction of metals from solution, while few attempts have been made to effect dry reductions by means of an electric current. Some time since Eugene H. Cowles and Alfred H. Cowles, of Cleveland, conceived the idea of obtaining a continuous high temperature on an extended scale by introducing into the path of an electric current some material that would afford the requisite resistance, thereby producing a corresponding increase in the temperature. After numerous experiments, coarsely pulverized carbon was selected as the best means for maintaining an invariable resistance, and at the same time as the most available substance for the reduction of oxides. When this material mixed with the oxide to be reduced was made a part of the electric circuit, enclosed in a fire-clay retort, and subjected to the action of a current from a powerful dynamo, not only was the oxide reduced,

* Ann Arbor Meeting, August 28, 1885.

but the temperature increased to such an extent that the whole interior of the retort fused completely. In other experiments lumps of lime, sand, and corundum were fused, with a reduction of the corresponding metal; on cooling, the lime formed large, well-defined crystals, the corundum beautiful red-green and blue octahedral crystals. Following up these results with the assistance of Prof. Mabery, who became interested at this stage, it was soon found that the intense heat thus produced could be utilized for the reduction of oxides in large quantities, and experiments were next tried on a large scale with the current from a fifty horse-power dynamo. For the protection of the walls of the furnace, which were of fire-brick, a mixture of ore and coarsely pulverized gas carbon was made a central core, and was surrounded on the side and bottom by fine charcoal, the current following the lesser resistance of the core from carbon electrodes inserted in the ends of the furnace in contact with the core. The furnace was charged by first filling it with charcoal, making a trough in the centre, and filling this with the ore mixture, the whole being covered with a layer of coarse charcoal. The furnace was closed on top with fire-brick slabs containing two or three holes for the escape of the gaseous products of the reduction, and the whole furnace was made air tight by luting with fire clay. Within a few minutes after starting the dynamo, a stream of carbonic oxide issued through the openings, burning usually with a flame eighteen inches high. The time required for complete reduction was ordinarily about an hour. Experience has already shown that aluminium, silicon, boron, manganese, sodium, and potassium can be reduced from their oxides with ease. In fact, there is no oxide that can withstand the temperature attainable in this furnace. Charcoal is changed to graphite; does this indicate fusion? As to what can be accomplished by converting enormous electrical energy into heat within narrow limits it can only be said that it opens the way into an extensive field of pure and applied chemistry. It is not difficult to conceive of temperature limited only by the power of carbon to resist fusion.

"Since the motive power is the chief expense in accomplishing reductions by this method, its commercial success is closely connected with obtaining power cheaply. Realizing the importance of this point, Messrs. Cowles have purchased at Lockport, N. Y., a water-power where they can utilize 1200 horse-power. An important feature in the use of these furnaces from a commercial standpoint is the slight technical skill required in their manipulation. The four furnaces operated in the experimental laboratory at Cleveland are in charge of two young men, who six months ago knew absolutely nothing of electricity. The products at present manufactured are the various grades of aluminium bronze, made from a rich furnace product obtained by adding copper to the charge of ore. Aluminium silver is also made; and a boron bronze may be prepared by the reduction of boracic acid in contact with copper, while silicon bronze is made by reducing silica in contact with copper. As commercial results may be mentioned the production in the experimental laboratory, which averages 50 pounds of 10 per cent. aluminium bronze daily, which can be supplied to the trade in large quantities on the basis of $5 per pound for the aluminium contained, the lowest market quotation of aluminium being now $15 per pound."

Dr. T. Sterry Hunt has written and read several papers on this furnace and process, and we extract from them anything not mentioned in Prof. Mabery's paper.

The following paper was read before the Am. Ins. of Mining Engineers by Dr. T. Sterry Hunt, of Montreal:—*

"The application of electricity in the extraction of metals has hitherto been chiefly confined to the electrolysis of dissolved or fused compounds by various methods. The power of electric currents to generate intense heat in their passage through a resisting medium has long been known, and the late Sir Wm. Siemens thereby succeeded in melting

* Halifax Meeting, Sept. 16, 1885.

considerable quantities of steel. Messrs. Cowles took a new step in the metallurgic art by making the heat thus produced a means of reducing, in presence of carbon, the oxides not only of the alkali metals, but of calcium, magnesium, manganese, aluminium, silicon, and boron, with an ease which permits the production of these elements and their alloys with copper and other metals on a commercial scale.

"If alumina, in the form of granular corundum, is mixed with the carbon in the electric path, aluminium is rapidly liberated, being in part carried off with the escaping gas and in part condensed in the upper layer of charcoal. In this way are obtained considerable masses of nearly pure aluminium, and others of a crystalline compound of the metal with carbon. When, however, some granular copper is placed with the corundum, an alloy of aluminium and copper is obtained, which is probably formed in the overlying stratum, but at the close of the operation is found in fused masses below. In this way there is obtained, after the current has passed an hour and a half through the furnace, four or five pounds of an alloy containing 15 to 20 per cent. of aluminium and free from iron. On substituting this alloy for the copper in a second operation, an alloy with over 30 per cent. aluminium is obtained. The difficulties in the way of gathering together the reduced metal without the aid of copper promise to be overcome at an early day, so that we may expect a cheap production of such alloys and of the pure metal. The present plant at Cleveland is but an experimental one, and has been in operation only a few months. The company will soon put in operation at Lockport a 125 horse-power dynamo, and nine more of equal power will be added, permitting the establishment of the electric furnace on a large scale."

Paper read before the National Academy of Science by Dr. Hunt:—*

"Dr. Hunt showed some alloys of aluminium with carbon and silicon, and a peculiar alloy believed to consist entirely of aluminium and nitrogen. As yet, the pure metal has only been produced direct from the furnace in small lumps, but it may be obtained by melting an alloy of aluminium and tin with lead, when the latter takes up the tin and separates from the aluminium, sinking beneath it. Or, we get aluminium by subliming either its alloy with carbon or with copper, when the pure aluminium is carried over. The maximum amount of aluminium which copper can tolerate is 10 per cent., until we approach the other end of the scale, when alloys with 70 to 80 per cent. of aluminium, or more, give valuable workable alloys. In the early experiments with the Cowles furnace, an engine of 30 horse-power running a dynamo yielded a daily output of 50 pounds of

* Washington Meeting, April 30, 1886.

10 per cent. aluminium bronze. Brush has now constructed an engine running 900 revolutions per minute, which for every 35 horse-power developed reduces one pound of the alloy per hour. The expense of working is now covered by one-half cent per horse-power per hour; thus the cost of the alloy is about 17 cents per pound. Within the past week, the gases given off by the furnace have been analyzed. In the first part of the process it is found that a large amount of nitrogen is given off, showing that air leaks into the furnace. After an hour and a half this gas is much diminished. The Cowles at first used moist carbon for packing, but have now overcome the necessity of dampening it, thereby saving the waste of heat in driving out the water."

The latest and most complete description of the process is a paper read by Mr. W. P. Thompson before the Liverpool Section of the Society of Chemical Industry.* Mr. Thompson has been Cowles Bros.' agent in taking out their patents in England. The paper is as follows:—

"That this invention is a new departure will be acknowledged by every one when they learn that chromium, titanium, silicon, aluminium, calcium, and the other alkaline earth metals are obtained by direct reduction of their oxides by carbon—till

* Jrnl. of the Soc. of Chem. Industry, April 29, 1886.

a year ago almost universally considered a practical impossibility.

"Conduction of the current of the large dynamo to the furnace and back is accomplished by a complete metallic circuit, except where it is broken by the interposition of the carbon electrodes and the mass of pulverized carbon in which the reduction takes place. The circuit is of 13 copper wires, each 0.3 inch in diameter. There is likewise in the circuit an ampère meter, or ammeter, through whose helix the whole current flows, indicating the total strength of the current being used. This is an important element in the management of the furnace, for, by the position of the finger on the dial, the furnace attendant can tell to a nicety what is being done by the current in the furnace. Between the ammeter and the furnace is a resistance coil of German silver kept in water, throwing more or less resistance into the circuit as desired. This is a safety appliance used in changing the current from one furnace to another, or to choke off the current before breaking it by a switch.

"The furnace (see Figs. 12, 13, 14) is simply a rectangular box, A, one foot wide, five feet long inside, and fifteen inches deep, made of firebrick. From the opposite ends through the pipes BB the two electrodes CC pass. The electrodes are immense electric-light carbons three inches in diameter and thirty inches long. If larger electrodes are required, a series this size must be used

REDUCTION BY OTHER AGENTS THAN SODIUM.

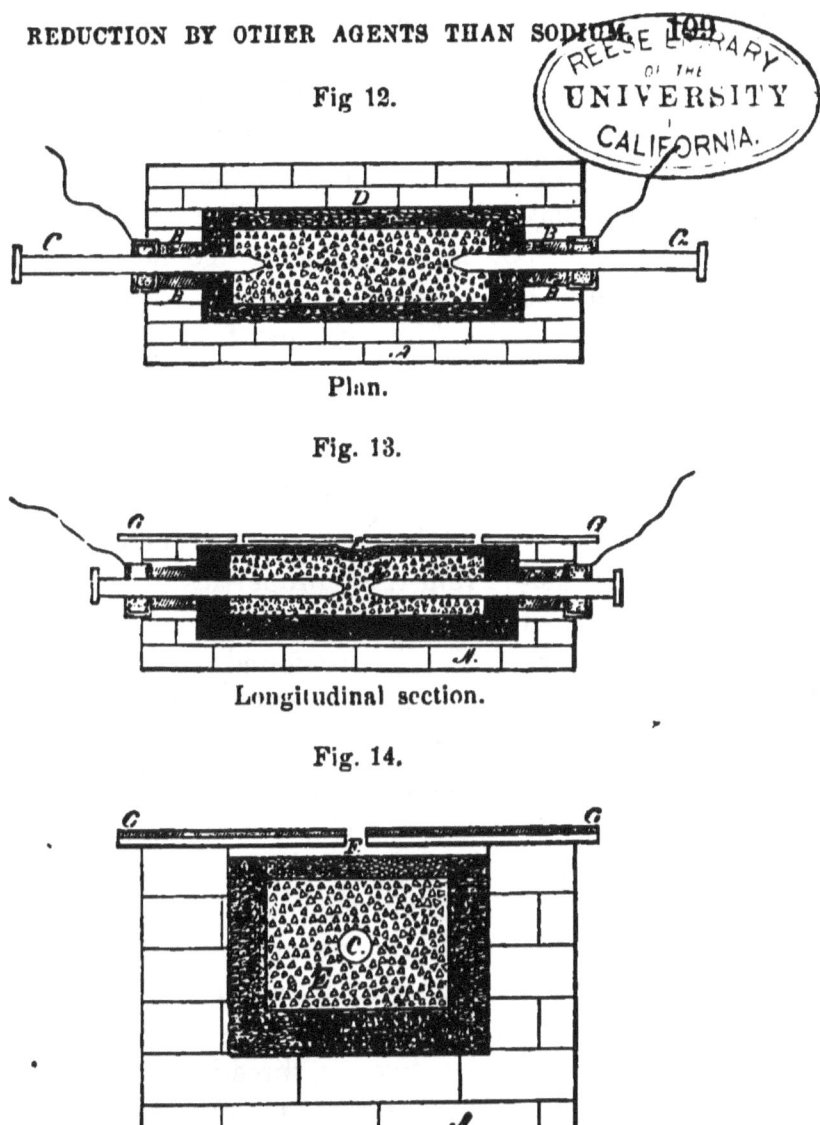

Fig 12.

Plan.

Fig. 13.

Longitudinal section.

Fig. 14.

Transverse section.

instead, as so far all attempts to make larger carbons that will not disintegrate on becoming incandescent have failed. The ends of the carbons are placed within a few inches of each other in the middle of the furnace, and the resistance coil and ammeter are placed in the circuit. The ammeter registers 50 to 2000 ampères. These connections made, the furnace is ready for charging.

"The walls of the furnace must first be protected, or the intense heat would melt the fire brick. The question arose, what would be the best substance to line the walls? Finely powdered charcoal is a poor conductor of electricity, is considered infusible and the best non-conductor of heat of all solids. From these properties it would seem the best material. As long as air is excluded it will not burn. But it is found that after using pure charcoal a few times it becomes valueless; it retains its woody structure, as is shown in larger pieces, but is changed to graphite, a good conductor of electricity, and thereby tends to diffuse the current through the lining, heating it and the walls. The fine charcoal is therefore washed in a solution of lime-water, and after drying, each particle is insulated by a fine coating of lime. The bottom of the furnace is now filled with this lining about two or three inches deep. A sheet-iron gauge is then placed along the sides of the electrodes, leaving about two inches between them and the side walls, in which space more of the charcoal is placed. The charge E, consisting of about 25 pounds of alumina, in its native form as corundum, 12 pounds of charcoal and carbon, and 50 pounds of granulated copper, is now placed within the gauge and spread around the electrodes to within a foot of each end of the furnace. In place of granulated copper, a series of short copper wires or bars can be placed parallel to each other and transverse to the furnace, among the alumina and carbon, it being found that where grains are used they sometimes fuse together in such a way as to short-circuit the current. After this, a bed of charcoal, F, the granules of which vary in size from a chestnut to a hickory, is spread over all, and the gauge drawn out. This coarse bed of charcoal above the charge allows free escape of the carbonic oxide generated in the reduction. The charge being in place, an iron top, G, lined with fire-brick, is placed over the whole furnace and the crevices luted to prevent access of air. The brick of the walls insulate the cover from the current.

"Now that the furnace is charged and the cover luted down, it is started. The ends of the electrodes were in the beginning placed close together, as shown in the longitudinal section, and for this cause the internal resistance of the furnace may be too low for the dynamo, and cause a short circuit. The operator, therefore, puts sufficient resistance into the circuit, and by watching the ammeter and now and then moving one of the electrodes out a trifle, he can prevent undue short circuiting in the beginning of the operation. In about ten minutes, the copper between the electrodes has been melted and the latter are moved far enough apart so that the current becomes steady. The current is now increased till 1300 ampères are going through, driven by 50 volts. Carbonic oxide has already commenced to escape through the two orifices in the top, where it burns with a white flame. By slight movements outwards of the electrodes during the coming five hours, the internal resistance in the furnace is kept constant, and at the same time all the different parts of the charge are brought in turn into the zone of reduction. At the close of the run the electrodes are in the position shown in the plan, the furnace is shut down by placing a resistance in the circuit and then the current is switched into another furnace charged in a similar manner. It is found that the product is larger if the carbons are inclined at angles of 30° to the horizontal plane.

"This regulating of the furnace by hand is rather costly and unsatisfactory. Several experiments have therefore been tried to make it self-regulating, and on January 26, 1886, a British patent was applied for by Cowles Bros., covering an arrangement for operating the electrodes by means of a shunt circuit, electro-magnet, and vibrating armature. Moreover, if the electrodes were drawn back and exposed to the air in their highly heated state, they would be rapidly wasted away. To obviate this, Messrs. Cowles place what may be called a stuffing box around them, consisting of a copper box filled with copper shot. The wires are attached to the boxes instead of the electrodes. The hot electrodes as they emerge from the furnace first encounter the shot, which rapidly carry off the heat, and by the time they emerge from the box they are too cool to be oxidized by contact with the air.

"Ninety horse-power have been pumped into the furnace for five hours. At the beginning of the operation the copper first melted in the centre of the furnace. There was no escape for the heat continually generated, and the temperature increased until the refractory corundum melted, and being surrounded on all sides by carbon gave up its oxygen. This oxygen, uniting with the carbon to form carbonic oxide, has generated heat which certainly aids in the process. The copper has had nothing to do with the reaction, as it will take place in its absence. Whether the reaction is due to the intense heat or to electric action it is difficult to say. If it be electric, it is Messrs. Cowles's impression that we have here a case where electrolysis can be accomplished by an alternating current, although it has not been tried as yet. Were the copper absent, the aluminium set free would now absorb carbon and become a yellow, crystalline carbide of aluminium; but, instead of that, the copper has become a boiling, seething mass, and the bubblings of its vapors may distinctly be heard. The vapors probably rise an inch or two, condense and fall back, carrying with them the freed aluminium. This continues till the current is taken off the furnace, when we have the copper charged with 15 to 30 per cent., and in some cases as high as 40 per cent. of its weight of aluminium, and a little silicon. After cooling the furnace this rich alloy is removed. A valuable property of the fine charcoal is that the metal does not spread and run through its interstices, but remains as a liquid mass surrounded below and on the sides by fine charcoal which sustains it just as flour or other fine dust will sustain drops of water for considerable periods, without allowing them to sink in. The alloy is white and brittle. This metal is then melted in an ordinary crucible furnace, poured into large ingots, the amount of aluminium in it determined by analysis, again melted, and the requisite amount of copper added to make the bronze desired.

"Two runs produce in ten hours' average work 100 pounds of white metal, from which it is estimated that Cowles Bros., at Lockport, are producing aluminium in its alloys at a cost of about forty cents per pound. The Cowles Co. will shortly have 1200 horse-power furnaces. With a larger furnace there is no reason why it should not be made to run continuously like the ordinary blast furnace.

"In place of the copper any non-volatile metal may be used as a condenser to unite with any metal it may be desired to reduce, provided, of course, that the two metals are of such a nature that they will unite at this high temperature. In this way aluminium may be alloyed with iron, nickel-silver, tin, or cobalt. Messrs. Cowles have made alloys containing 50 Al and 50 Fe, 30 Al, and 70 Cu, 25 Al and 75 Ni. Silicon or boron or other rare metals may be combined in the same way, or tertiary alloys may be produced; as, for instance, where fire clay is reduced in presence of copper we obtain an alloy of aluminium, silicon, and copper. This alloy is white and brittle if it contains over ten per cent. of aluminium and silicon together. With from two to six per cent. of these two, in equal proportions, the alloy is stronger than gun-metal, has great toughness, does not oxidize when heated in the air, and has a fine gold color. I hear to-day that an aluminium-silicon bronze wire made by Cowles has shown a tensile strength of 200,000 pounds, hitherto unprecedented in any metal.

"As to the ores of aluminium. For Mitis castings, where iron and silicon are not prejudicial, beauxite or various clays may be used to advantage. For bronze making, alumina containing silica in considerable quantities is as available as the pure earth and is indeed superior to it." To manufacture pure aluminium, pure alumina is necessary. Cowles Bros. use corundum obtained from Northern Georgia. (See p. 49.)

Reduction by Iron.

The statement has been made* that aluminium sulphide, Al^2S^3, is to be obtained from powdered cryolite by treating it with water, which dissolves out sodium fluoride, NaF, and the residual Al^2F^6 being calcined with sulphide of lime, CaS, there results Al^2S^3 and CaF^2. The Al^2S^3 is then decomposed by heating to redness with iron turnings.

According to a patent given to F. Lauterborn,† Germany, Aug. 14, 1880, if pulverized cryolite is boiled with water, NaF is set free and Al^2F^6 remains. Likewise, calcium fluoride, CaF^2, boiled with Al^2Cl^6 gives $CaCl^2$ and Al^2F^6. The aluminium fluoride by heating with sulphide of lime will be converted into Al^2S^3. Finally, the Al^2S^3, by heating red hot with iron gives, it is claimed, metallic aluminium.

The above are all the details of this process to be found. See in the Appendix an experiment on thus decomposing cryolite.

II. Niewerth‡ has patented the following process: "Ferro-silicum is mixed with Al^2F^6 in proper proportions and the mixture submitted to a suitable red or melting heat by which the charge is decomposed into volatile silicon fluoride, SiF^4, iron, and aluminium, the two latter forming an alloy.

* Chemical News, 1860.
† Dingler, 242, p. 70.
‡ Sci. Am. Suppl. Nov. 17, 1883.

In order to obtain the valuable alloy of aluminium and copper from this iron-aluminium alloy, the latter is melted with metallic copper, which will then by reason of greater affinity unite with the aluminium, while the iron will retain but an insignificant amount of it. On cooling the bath, the bronze and iron separate in such a manner that they can readily be kept apart. In place of pure Al^2F^6, cryolite may advantageously be employed, or Al^2Cl^6 may also be used, in which case silicon chloride volatilizes instead of the fluoride. Or, again, pure silicon may be used with Al^2F^6, cryolite, or Al^2Cl^6, in which case pure aluminium is obtained."

Preparation of Aluminium and Sodium im the Bessemer Converter.

According to the experiments of Mr. W. P. Thompson,* sodium and aluminium may be advantageously prepared by means of a Bessemer converter. The same process it seems should serve equally well for the preparation of the other difficultly reducible metals, such as calcium, strontium, barium, magnesium, etc.

· Mr. W. P. Thompson† has taken out a patent in England‡ for the manufacture of aluminium and similar metals, which is carried out as follows: The inventor employs as a reducing agent iron,

---

* Bull. de la Soc. Chem. de Paris, 1880, xxiv. 128.
† Idem. p. 719.   ‡ Mar. 27, 1879. No. 2101.

either alone or conjointly with carbon or hydrogen. The operation is effected in an apparatus similar to a Bessemer converter, divided into two compartments. In one of these compartments is placed melted iron, or an alloy of iron, which is made to run into the second by turning the converter. This last compartment has two tuyeres, one of which serves to introduce hydrogen, while by the other is introduced either Al^2Cl^6, Al^2F^6, $Al^2Cl^6.2NaCl$, or $Al^2F^6.6NaF$, in liquid or gaseous state. In presence of the hydrogen the iron takes up chlorine or fluorine, chloride or fluoride of iron is disengaged, and aluminium mixed with carbon remains as a residue. Then this mixture of iron, aluminium, and carbon is returned to the other compartment where the carbon is burnt out by means of a current of air. The mass being then returned to the chamber of reduction, the operation described is repeated. When almost all the iron has been consumed, the reduction is terminated by hydrogen alone. There is thus obtained an alloy of iron and aluminium. (The preparation of sodium does not require the intervention of hydrogen. A mixture of iron with an excess of carbon and caustic soda, $NaOH$, is heated in the converter, when the sodium distils off.* When all the carbon has been burnt, the iron remaining as a residue may be converted into Bessemer steel. As iron forms an alloy with potassium, the method

* Compare with p. 141.

would scarcely serve for the production of that metal.) To obtain the pure aluminium, sodium is first prepared by the process indicated, the chloride or fluoride of aluminium is introduced into the apparatus in the other chamber, when the metal is reduced by the vapor of sodium. The chambers ought to be slightly inclined, and an agitator favors the reaction. The inventor intends to apply his process to the manufacture of magnesium, strontium, calcium, and barium.

Calvert and Johnson[*] made experiments on the reduction of aluminium by iron, and the production thereby of iron-aluminium alloys. We give the report in their own words:—

"We shall not describe all the fruitless efforts we made, but confine ourselves only to those which gave satisfactory results. The first alloy we obtained was by heating to a white heat for two hours the following mixture:—

8 equivalents of Al^2Cl^6	. .	.	1076 parts.
40 "	" iron filings	. .	1120 "
8 "	" lime	. . .	224 "

"The lime was added to the mixture with the view of removing the chlorine from the Al^2Cl^6, so as to liberate the metal and form fusible calcium chloride, $CaCl^2$. Subtracting the lime from the above proportion, we ought to have obtained an alloy having the composition of 1 Equivalent Al

[*] Phil. Mag., 1855, x. 240.

to 5 Equivalents of Fe, or with 9.09 per cent. aluminium. The alloy we obtained contained 12 per cent., which leads to the formula $AlFe^4$. This alloy, it will be noticed, has an analogous composition to the one we made of iron and potassium, and like it was extremely hard, and rusted when exposed to a damp atmosphere. Still it could be forged and welded. We obtained a similar alloy by adding to the above mixture some very finely pulverized charcoal and subjecting it to a high heat in a forge furnace for two hours. This alloy gave on analysis 12.09 per cent.* But, in the mass of $CaCl^2$ and carbon remaining in the crucible there was a large amount of globules varying in size from a pin head to a pea, as white as silver and extremely hard, which did not rust in the air or in hyponitric fumes. Its analysis gave 24.55 per cent. aluminium; the formula Al^2Fe^3 would give 25 per cent. Therefore this alloy has the same composition as Al^2O^3, iron replacing oxygen. We treated these globules with weak sulphuric acid, which removed the iron and left the aluminium, the globules retaining their form, and the metal thus obtained had all the properties of the pure aluminium.

"We have made trials with the following mixture, but, although they have yielded results, still they are not sufficiently satisfactory to describe in

* In the original paper it is given as 12.09 per cent. iron. The inference is unavoidable that this was a misprint, but it is not corrected in the Errata at the end of the volume.

this paper, which is the first of a series we intend publishing on alloys. This mixture was:—

Kaolin	1750 parts.
NaCl	1200 "
Fe	875 "

"From this we obtained a metallic mass and a few globules which we have not yet analyzed."

Fremy: Alloys of aluminium and iron have been prepared by Benzon by calcining a mixture of alumina, carbon, and iron or Fe^2O^3. (See p. 214.)

Watts: E. L. Benzon[*] reduces aluminium by heating alumina with the oxide of another metal, as of copper, iron, zinc, or a mixture of alumina with carbon and the other metal in a free state, the materials being all finely divided and mixed in atomic proportions, or rather with the carbon slightly in excess.

M. Evrard,[†] in order to make aluminium bronze, makes use of an aluminous pig iron. (It is not stated how this aluminous pig iron is made.) This is slowly heated to fusion, and copper is added to the melted mass. Aluminium, having more affinity for copper than for iron, abandons the latter and combines with the copper. After the entire mass has been well stirred, it is allowed to cool slowly so as to permit the bronze, which is heavier than iron, to find its way to the bottom

---

[*] Eng. Pat., 1858, No. 2753.
[†] Annales du Genie Civil, Mars, 1867, p. 189.

of the crucible. M. Evrard makes silicon bronze in the same way by using siliceous iron.

'Eng. and Mining Journal,' May 15, 1886 : " The iron-aluminium alloy used in the Mitis process, we are informed by Mr. Ostberg, is made in Sweden by the addition of clays in iron smelting, a patented process producing alloys with 7 to 8 per cent. aluminium very cheaply. Mr. Ostberg adds that he purchased a small quantity of Cowles Bros.' alloy, which gave rise to our previous unqualified statement that he used Cowles' alloys." (See ' Mitis Castings,' Part XI.).

REDUCTION WITH COPPER.

Calvert and Johnson[*] obtained copper alloyed with aluminium by recourse to a similar chemical reaction to that employed to get their iron-aluminium alloy. Their mixture was composed of—

20 equivalents of Cu	. . . .	640 parts.
8 " " Al^2Cl^6	. . . .	1076 "
10 " " CaO	. . . .	280 "

"We mixed these substances intimately together, and after having subjected them to a high heat for one hour we found at the bottom of the crucible a melted mass covered with cuprous chloride, Cu^2Cl^2, and in this mass small globules, which on analysis contained 8.47 per cent. aluminium, corresponding to the formula—

[*] Phil. Mag. 1855, x. 242.

5 equivalents of Cu . . 160 . . 91.96 per cent.
1   "   " Al . . 14 . . 8.04   "
                      ———   ———
                      174   100.00

"We made another mixture of Al^2Cl^6 and copper in the same proportions as above, but left out the lime. We obtained an alloy in this case also, which contained 12.82 per cent. aluminium, corresponding to the formula—

3 equivalents of Cu . . 96 . . 87.27 per cent.
1   "   " Al . . 14 . . 12.73   "
                     ———   ———
                     110   100.00

Kerl and Stohman give the following account of Benzon's process: "Benzon[*] has patented the reduction of aluminium with copper, forming an aluminium-copper alloy. He mixes copper, or oxidized copper, or cupric oxide, in the finest possible state, with fine, powdered, pure alumina and charcoal, preferably animal charcoal. The alumina and copper or copper oxide are mixed in equivalent proportions, but an excess of charcoal is used. The mixture is put in a crucible such as is used for melting cast steel, which is lined inside with charcoal. The charge is covered with charcoal, and the crucible subjected first to a temperature near the melting point of copper, until the alumina is reduced, and then the heat is raised high enough to melt down the alloy. In this way can be obtained a succession of alloys, whose hard-

[*] Eng. Pat. 1858, No. 2703.

ness and other qualities depend on the percentage of aluminium in them. In order to obtain alloys of a certain composition, it is best to produce first an alloy of the highest attainable content of aluminium, to analyze it, and then melt it with the required quantity of copper. The same process can be used for the reduction of alumina with iron or Fe^2O^3, only the carbon must in this case be in greater excess, and a stronger heat kept up longer must be used than when producing the copper-aluminium alloy. In contact with Fe^2O^3 the alumina is more easily reduced than with metallic iron."

Kerl and Stohman remark that were these methods practicable, then at once there is the possibility of producing copper-aluminium alloys at a low price, and, on the other hand, of easily producing pure aluminium from the iron alloy. According to researches conducted in the laboratories at Zurich and Augsburg, it was found that the melted-down copper contained either no aluminium or at most a trace. (See Appendix.)

Aluminium-bronze is also made by Mr. Evrard's process given on p. 211.

REDUCTION BY ZINC.

M. Dullo* observes that the double chloride of aluminium and sodium, which he makes directly

---

* Bull. de la Soc. Chem. 1860, v. 472.

from clay, may be reduced by zinc. He says, "The reduction by zinc presents no difficulties, but it is less easy than with sodium. An excess of zinc should be employed, which may be got rid of afterwards by distillation. The metal thus prepared possesses all the characteristics and all the properties of that obtained from beauxite with sodium."

M. N. Basset,[*] a chemist in Paris, has recently patented a new process for obtaining aluminium. If the statements are correct they are of great value. The paper is as follows: All the metalloids and the metals which form by double decomposition proto-chlorides or sesqui-chlorides more fusible or more soluble than Al^2Cl^6 may reduce Al^2Cl^6 or even $Al^2Cl^6.2NaCl$. Thus, As, Bi, Cu, Zn, Sb, Hg, or even Sn, or amalgam of Zn, Sb, or Sn may be employed to reduce the single or double chloride. The author employs zinc in preference to the others in consequence of its low price, the facility of its employment, its volatility, and the property which it has of metallizing easily the aluminium as it is set free. When metallic zinc is put in the presence of $Al^2Cl^6.2NaCl$ at 250 to 300°, zinc chloride, $ZnCl^2$, is formed and aluminium is set free. This dissolves in the zinc present in excess, the $ZnCl^2$ combines with the NaCl, and the mass becomes little by little pasty, then solid, while the alloy

[*] Le Genie Industriel, 1862, p. 152.

remains fluid. If the heat is now raised, the mass melts anew, the zinc reduces a new portion of the double chloride, and the excess of zinc enriches itself in aluminium proportionately. These facts constitute the basis of the following general process: One equivalent of Al^2Cl^6 is melted, two of NaCl added, and when the vapors of hydrochloric acid are dissipated, four equivalents of zinc, in powder or grain, is introduced. The zinc melts rapidly, and by agitation the mass of chloride thickens and solidifies. The mass is now composed of Al^2Cl^6, NaCl, and $ZnCl^2$, and remains in a pasty condition on top of the fluid zinc containing aluminium. This pasty mass is removed, piled up in a crucible or in a furnace, and bars of the fluid alloy of zinc and aluminium obtained from a previous operation are placed on top of it. This is gradually heated to bright redness, and kept there for an hour. The melted mass is then stirred with a rake and poured out. It is an alloy of the two metals in pretty nearly equal proportions. This alloy, melted with some chloride from the first operation furnishes aluminium containing only a small per cent. of zinc, which disappears by a new fusion under chloride mixed with a little Al^2F^6, providing the temperature is raised to a white heat and maintained till the cessation of the vapors of zinc, air being excluded.

The metal is pure if the zinc employed contained no foreign materials or metals. It is melted and cast into ingots. In case the zinc contains iron, or even if the Al^2Cl^6 contains some, the metallic product of the second operation may be treated with dilute sulphuric acid to remove it. The insoluble residue is washed and melted layer by layer with fluorspar or cryolite and a small quantity of Al^2Cl^6.-$2NaCl$, intended solely to help the fusion."

Mr. Wedding* makes the following remarks on this process:—

"It is some time since Mr. Basset established the possibility of replacing sodium by zinc in the manufacture of aluminium. Operating on Al^2Cl^6.-$2NaCl$ with granulated zinc, the reduction takes place towards 300°. The reduced aluminium dissolves in the excess of zinc, while the $ZnCl^2$ formed combines with the NaCl, forming a pasty mass if the heat is not raised. Under the action of heat the alloy enriches itself in aluminium, because the zinc volatilizes. The zinc retained by this alloy is completely eliminated by fusion with $Al^2Cl^6.2NaCl$ and a little fluorspar. The temperature ought to be pushed at last to a white heat, and maintained till no vapor of zinc escapes, air being excluded during the operation. These results I have confirmed, having submitted the experiments of Mr. Basset to an attentive examination, and I recommend its use. However, the process demands very much precaution because of the high temperature

* Journal de Pharm. [4] iii. p. 155 (1866).

which it necessitates. Another chemist, Mr. Specht, even in 1860 decomposed Al^2Cl^6 by zinc, and has the same report to make—that he thinks the process will be some time advantageously practised on a large scale."

However, this method has not succeeded in being established in practice, probably on account of the high temperature which is necessary to drive off the zinc, in which operation some aluminium is lost.

Kagensbusch,* in Leeds, makes the singular proposition to melt clay with fluxes; then, by adding zinc or lead, to decompose it by an electrical current and isolate an aluminium-zinc or aluminium-lead alloy, from which the zinc may be volatilized or the lead cupelled.

Mr. Fred. J. Seymour† patents the reduction of aluminium by zinc, and makes the following claim: An improvement in extracting aluminium from aluminous earths and ores by mixing them with an ore of zinc, carboniferous material and a flux, and subjecting the mixture to heat in a closed retort, whereby the zinc is liberated, is caused to assist in bringing or casting down the aluminium in a metallic state, and an alloy of aluminium and zinc is obtained.

The only information outside of the patent claims

* Eng. Pat., 1872, No. 4811.
† U. S. Pat., No. 291,631, Jan. 8, 1884.

which I could find in regard to this process is contained in the following newspaper article, which, although wordy and indefinite, will have to be taken in the absence of a more precise account.

"Mr. F. J. Seymour,* a well-known practical metallurgist, late of Bridgeport, Conn., has, as the result of several years' study, succeeded in producing aluminium at a low cost, and by the novel furnace just designed asserts that he can extract the metal on a commercial basis in large quantities. Not to go into all the technical details, which are extremely interesting to metallurgists, it is sufficient to say that Mr. Seymour has discovered that the close affinity existing between aluminium and zinc can be utilized in vaporizing, capturing, and depositing the aluminium, the separation being effected by the aid of heat in a furnace, or rather a series of furnaces, of peculiar construction. The charge for each furnace is zinc ore 100 parts, koalin 50, carbon (either anthracite coal or its equivalent in hydrocarbon gas) 125, pearl ash or its equivalent 15, NaCl 10; all intimately mixed. The retorts are of steel, 36 inches long, 12 inches diameter, sides ⅞ inch thick. The heat necessary to produce the result is about 2500° F., or 1400° C. Properly handled, one furnace should make two charges in 24 to 30 hours. Four men can operate 50 retorts.

* Cleveland Letter to the 'New York Times,' April 14, 1884.

The number of retorts can be increased to several hundred in a single system. Capitalists are already interested in this new process, and the prospects are that operations on an extensive scale will soon follow. Independent investigations in the same line in this city have resulted in the recent incorporation of a company for the extraction of aluminium by electricity. Thus far the secret of the process has been strictly guarded, and no details can be given."

Mr. Seymour has quite recently taken out another patent, the claims of which are hardly reconcilable with those of the former patent. The claim is as follows:—

Patent to Fred. J. Seymour,* Wolcottville, Conn., assignor of one-half to Mr. Henry Brown, New York. The following is the claim: "The process of extracting aluminium from aluminous earths, consisting in subjecting such ore or earth with an ore of zinc, carbonaceous matter, and a flux, to heat, in a retort; wherein the oxides of aluminium and zinc are vaporized; collecting and condensing the vapors in a condenser, and afterwards subjecting the condensed product to heat with carbonaceous matter, substantially as herein described."

If Mr. Seymour can make a process work according to the details of the above extraordinary claim,

---

* U. S. Pat. No. 337,996, filed March, 1885, granted March 16, 1886.

he will certainly have a claim on the admiration of all scientific men. The idea of vaporizing the oxides of zinc and aluminium is certainly unique. I wrote to Mr. Seymour, asking for further details of his process, and if he was making any aluminium, but have received no further information than has already been given.

'The American Machinist,' August, 1886, contains the statement that the American Aluminium Company has been organized at Detroit with a capital stock of $2,500,000; to use the patents of Dr. Smith for the United States, Great Britain and France. I was informed by a gentleman in the aluminium industry that this company were to operate Mr. Seymour's zinc process.

Reduction by Lead.

According to the invention of Mr. A. E. Wilde,[*] of Notting Hill, lead or sulphide of lead, or a mixture of the two, is melted and in a molten state poured upon dried or burnt alum. The crucible in which the mass is contained is then placed in a furnace and heated, with suitable fluxes. The metal, when poured out of the crucible, will be found to contain aluminium. The aluminium and lead can be subsequently separated from each other by any known means, or the alloy or mixture of the two metals

[*] Sci. Am. Suppl., Aug. 11, 1877.

can be employed for the various useful purposes for which lead is more or less unsuited.

Kagensbusch's process, using lead, is described on p. 218 under the reduction by zinc.

REDUCTION BY MANGANESE.

W. Weldon,* of Burstow, Eng., claims to melt together cryolite with $CaCl^2$ or another non-metallic chloride or sulphide, and then to reduce the Al^2Cl^6 or Al^2S^3 produced with manganese, which he claims is even powerful enough to reduce sodium.

REDUCTION BY ELECTRICITY.

The reduction of Al^2Cl^6 or $Al^2Cl^6.2NaCl$ by sodium is the only process by which the pure metal is now made. However, many attempts have been made to isolate it by means of the electric current. The reduction may take place in either the dry or wet way. The reduction of fused $Al^2Cl^6.2NaCl$ by the battery was accidentally discovered simultaneously by Deville in France and Bunsen in Germany, in 1854, and is nothing else but an application of the process already announced by Bunsen of decomposing magnesium

* Eng. Pat., 1883, No. 97. Wagner's Jahresb., 1884.

chloride, $MgCl^2$, by the battery. Deville's account of the process is as follows:—*

"It appears to me impossible to obtain aluminium by the battery in aqueous solutions. I should believe this to be an impossibility if the brilliant experiments of Mr. Bunsen in the preparation of barium did not shake my convictions. Still I must say that all the processes of this description which have recently been published for the preparation of aluminium have failed to give me good results. To prepare the bath for decomposition in the dry way, I heated a mixture of 2 parts Al^2Cl^6 and 1 part NaCl, dry and pulverized, to about 200° in a porcelain capsule. They combine with disengagement of heat, and the resulting bath is very fluid. The apparatus which I use for the decomposition comprises a glazed porcelain crucible, which as a precaution is placed inside a larger one of clay. The whole is covered by a porcelain cover pierced by a slit to give passage to a large, thick leaf of platinum, which serves as the negative electrode; the lid has also a hole through which is introduced, fitting closely, a well-dried porous cylinder, the bottom of which is kept at some distance from the inside of the porcelain crucible. This porous vessel incloses a pencil of retort carbon, which serves as the positive electrode. Melted $Al^2Cl^6.2NaCl$ is poured into the porous jar and into the crucible so as to stand at the same

* Ann. de Chem. et de Phys. [3], 46, 452.

224 ALUMINIUM.

height in both vessels; the whole is heated just enough to keep the bath in fusion, and there is passed through it the current from several Bunsen cells, two cells being strictly sufficient. The annexed diagram shows the crucibles in section.

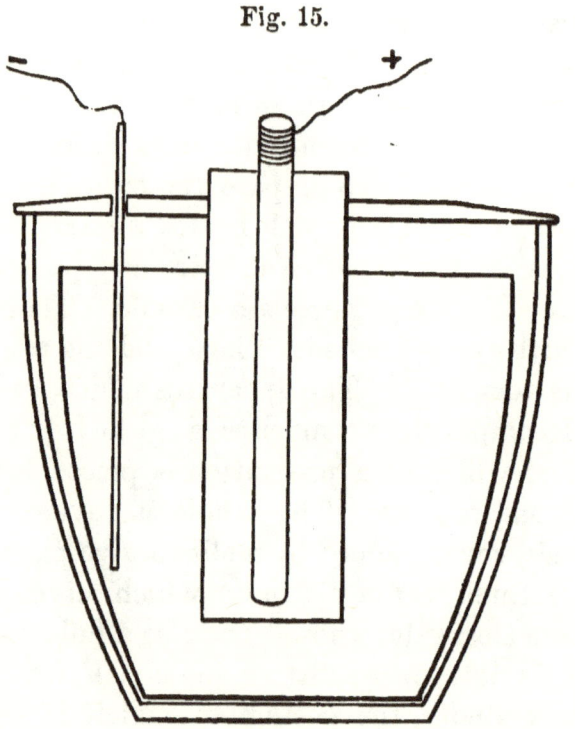

Fig. 15.

"The aluminium deposits with some NaCl on the platinum leaf; the chlorine, with a little Al^2Cl^6, is disengaged in the porous jar, and forms white fumes, which are prevented from rising by throwing into the jar from time to time some dry, pulverized NaCl. To collect the aluminium, the platinum leaf is removed when sufficiently charged with the saline and metallic deposit; after letting it cool the deposit is rubbed off and the leaf placed in its former position. The material thus detached, melted in a porcelain crucible, and after cooling washed with water, yields a gray, metallic powder, which is melted under a layer of $Al^2Cl^6.2NaCl$ and reunited into a button."

Bunsen* adopted a similar arrangement. The porcelain crucible containing the bath of Al^2Cl^6. $2NaCl$ kept in fusion was divided into two compartments in its upper part by a partition, in order to separate the chlorine liberated from the aluminium reduced. He made the two electrodes of retort carbon. To reunite the pulverulent aluminium, Bunsen melted it in a bath of $Al^2Cl^6.2NaCl$, continually throwing in enough NaCl to keep the temperature of the bath about the fusing point of silver.

Deville,† without being acquainted with Bunsen's investigations, employed the same arrangement, but he abandoned it because the retort carbon slowly disintegrated in the bath, and a considerable quantity of $Al^2Cl^6.2NaCl$ was lost by the higher heat necessary to reunite the globules of aluminium after the electrolysis. Deville also

* Pogg, 97, 648.
† Ann. de Phys. et de Chem. [3], 43, 27.

observed that by working at a higher temperature, as Bunsen has done, he obtained purer metal, but in less quantity. The effect of the high heat is that silicon chloride is formed and volatilizes, and the iron which would have been reduced with the aluminium is transformed to $FeCl^2$ by the Al^2Cl^6, and thus the aluminium is purified of silicon and iron.

Mierzinski makes the following practical remarks on the use of electricity in producing aluminium:—

"An important factor which we must notice in the present production of aluminium is the application of electricity. On all sides the greatest efforts are being made to apply electricity to chemical technology; in the future the importance of electricity will centre on its application to reducing metals. Even in the year 1807 Davy succeeded in decomposing caustic potash by means of a current from a 400 element Wollaston battery. But we now have magneto-electric and dynamo-electric machines which are much lighter and cheaper than they were in Davy's time. The application of electricity for producing metals also possesses the advantage not to be ignored, that a degree of heat may be attained with it such as cannot be reached by a blowpipe or regenerative gas furnace. The highest furnace temperature attainable is 2500 to 2800° C., but long before this point is reached the combustion becomes so languid that the loss of heat by radiation almost equals the production of heat by combustion, and hinders a further elevation of temperature. But in applying electricity, the degree of heat attainable is theoretically unlimited. A further advantage is that the smelting takes place in a perfectly neutral atmosphere, the whole operation goes on without much preparation and under the eyes of the operator. Finally, in ordinary furnaces the refractory material of the vessel must stand a higher heat than the substance in it, whereas by smelting in an electrical furnace the material to be fused has a higher temperature than the crucible itself.

" The manufacture of aluminium is effected now either by separating out the metal itself directly from the solutions of its salts or by reducing it with sodium. However, in spite of numerous attempts, sodium has not been replaced as a reducing agent. In the production of aluminium, the making of Al^2O^3 from beauxite costs 9.7 per cent., making the $Al^2Cl^6.2NaCl$ 33.4 per cent., and decomposing by sodium 56.9 per cent. of the whole cost. The attempt to reduce alumina directly by carbon Mr. W. Weldon considers as impossible because he could not produce the temperature required for the reaction to take place. Hence appears the great importance of utilizing the temperature attainable by the electric current. The separation of aluminium by electrolysis is now done only by the use of anhydrous $Al^2Cl^6.2NaCl$, melting at 200° C. The anodes are made of plates of alumina and carbon pressed together, having the conducting wire leading through their whole length in order to lessen the resistance as much as possible. The metal is obtained as a granular powder mixed with NaCl. Where possible, vessels of chalk or magnesia should be used, since aluminium takes up silicon from siliceous crucibles and becomes brittle."

There have been some improvements made in the form of apparatus over those used by Bunsen and Deville, designed to produce the metal on a commercial scale. The best one is that patented in Germany by Richard Grätzel.* He uses melting-pots of porcelain, alumina, or aluminium, which serve also as negative electrodes. A number of these are placed in one furnace. The following section shows the arrangement (Fig. 16). The positive electrode K can be made of a mixture of anhydrous alumina and carbon pressed into shape and ignited. A mixture of alumina and gas-tar answers very well; or it can even be made of gas-tar and gas-retort carbon. During the operation little pieces of carbon fall from it and would contaminate the bath, but are kept from doing so by the mantle G. This isolating vessel G is perforated around the lower part at g, so that the chlorine gas liberated at K may escape through the tube O', while reducing gases can be brought into the cruci-

* D. R. Pat. No. 26,962.

REDUCTION BY OTHER AGENTS THAN SODIUM. 229

ble by the tube O^2. To lessen the electrical resistance and to renew the bath of chloride or fluoride,

Fig. 16.

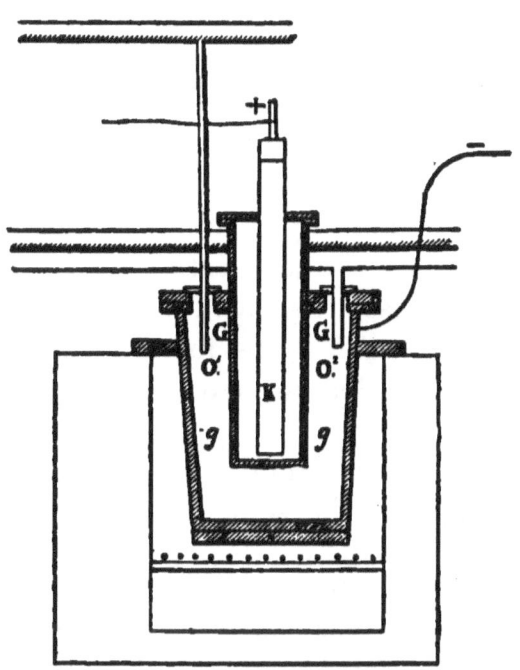

bars of carbon, alumina, or magnesia are placed inside the isolating mantle G.

This process is now being worked on a large scale in Germany, being also used for producing magnesium. There are works at Bremen and Hamburg.

M. Duvivier[*] states that by passing an electric

[*] The Chemist, Aug. 1854.

current from eighty Bunsen cells through a small piece of laminated disthene between two carbon points, the disthene melted entirely in two or three minutes, the elements which composed it were partly disunited by the power of the electric current, and some aluminium freed from its oxygen. Several globules of the metal separated, one of which was as white and as hard as silver.

Kagensbusch,* of Leeds, makes the singular proposition to melt clay with fluxes, then add zinc or a like metal, pass an electric current through the fused mass, isolating an alloy of aluminium and the metal, from which the foreign metal may be removed by distillation, sublimation, or cupellation.

Gaudin† reduces aluminium by a process to which he applies the somewhat doubtful title of economic. He melts together equal parts of cryolite and NaCl, and traverses the fused mass by an electric current. Fluorine is evolved at the positive pole, while aluminium accumulates at the negative.

Thus far we have given the methods based on electrolyzing fused salts. These seem to be the operations best suited to throwing down aluminium in mass. The electrolysis of aqueous solutions seems so far to have succeeded only in depositing very thin films of metal. We will now give the

* Eng. Pat. 1872, No. 4811.
† Moniteur Scientifique, xi. 62.

various methods proposed for electrolyzing aluminium in the wet way.

Messrs. Thomas and Tilly* coat metals with aluminium and its alloys by using, for depositing the pure metal, a solution of freshly precipitated alumina dissolved in boiling water containing potassium cyanide, or a solution of freshly calcined alum in aqueous potassium cyanide; also from several other liquids. Their patent covers the deposition of the alloys of aluminium with silver, tin, copper, iron, silver and copper, silver and tin, etc. etc.

M. Corbelli, of Florence,† deposits aluminium by electrolyzing a mixture of rock alum or sulphate of alumina with $CaCl^2$ or NaCl, in aqueous solution, the anode being mercury placed at the bottom of the solution and connected to the battery by an iron wire coated with insulating material and dipping its uncovered end into the mercury. The zinc cathode is immersed in the solution. Aluminium is deposited on the zinc, and the chlorine which is liberated at the anode unites with the mercury, forming calomel.

J. B. Thompson‡ reports that he has for over two years been depositing aluminium on iron, steel, and other metals, and also depositing aluminium

---

* Eng. Pat., 1855, No. 2756.
† Eng. Pat., 1858, No. 507.
‡ Chem. News, xxiv. 194.

bronze of various tints, but declines to state his process.

J. A. Jeancon* has patented a process for depositing aluminium from an aqueous solution of a double salt of aluminium and potassium of specific gravity 1.161; or from any solution of an aluminium salt, such as sulphate, nitrate, cyanide, etc., concentrated to 20° B. at 50° F. He uses a battery of four pairs of Smee's or three Bunsen's cells, with elements arranged for intensity, and electrolyses the solutions at 140° F. The first solution will decompose without an aluminium anode, but the others require such an anode on the negative pole. The solution must be acidulated slightly with acid corresponding to the salt used, the temperature being kept at 140° F. constantly.

M. A. Bertrand† states that he deposited aluminium on a plate of copper from a solution of double chloride of aluminium and ammonia, by using a strong current, and the deposit was capable of receiving a brilliant polish.

C. Winkler‡ states that he has spent much time and tried all methods so far proposed, and comes to the conclusion that aluminium cannot be deposited by electro-deposition in the wet way.

---

* Annual Record of Science and Industry, 1875.
† Chem. News, xxxiv. 227.
‡ Journal of the Chem. Soc., x. 1134.

Sprague* also states his inability to deposit aluminium electrically from solution.

M. L. Senet† electrolyzes a saturated solution of $Al^2(SO^4)^3$, separated by a porous septum from a solution of NaCl. A current is used of four ampères. The double chloride, $Al^2Cl^6.2NaCl$, is formed, then decomposed, and the aluminium liberated deposited on the negative electrode.

Gerhard and Smith‡ patented a process for depositing electrically aluminium and its alloys.

John Braun§ decomposes a solution of alum, of specific gravity 1.03 to 1.07, at the usual temperature, using an insoluble anode. In the course of the operation, the sulphuric acid set free is neutralized by the continual addition of alkali; and, afterwards, to avoid the precipitation of alumina, a non-volatile organic acid is added to the solution.

Moses G. Farmer‖ has patented an apparatus for obtaining aluminium electrically consisting of a series of conducting cells in the form of ladles, each ladle having a handle of conducting material extending upwards above the bowl of the next succeeding ladle; each ladle can be heated separately from the rest; the anodes are hung in the ladles, being suspended from the handles of the

---

* Sprague's Electricity, p. 309.
† Cosmos les Mondes, Aug. 10, 1885.
‡ Eng. Pat., 1884, No. 16,653.
§ German Pat., No. 28,760.
‖ U. S. Pat., No. 315,266, Apr. 1885.

preceding ladles, the ladles themselves being the cathodes.

Mierzinski says that the deposition of aluminium from an aqueous solution of its salt has not yet been accomplished, and declares Gore to have been in error when he stated that he had covered copper with a film of aluminium by using a feeble current and a solution of Al^2Cl^6 in water.

Several years ago, the writer was in Mr. Frishmuth's works, in Philadelphia, and observed that he was then doing a large amount of plating, depositing an alloy of aluminium and nickel. Nickel plating is known to be very hard and lasting, but it has a dark-bluish color, not agreeable to many. The presence of aluminium with it whitens it so that the plating is a very close imitation of silver, and wears much better than silver plating. He was depositing with a twenty horse-power dynamo. The articles were previously cleaned in a hot potash solution, and then hung in the plating bath. I do not know the composition of his solution, he keeps that secret, but it was green and strongly ammoniacal.

PART XI.

WORKING IN ALUMINIUM.

MELTING ALUMINIUM.

DEVILLE: To melt aluminium it is necessary to use an ordinary earthen crucible and no flux. Fluxes are always useless and almost always harmful. The extraordinary chemical properties of the metal are the cause of this; it attacks very actively borax or glass with which one might cover it to prevent its oxidation. Fortunately this oxidation does not take place even at a high temperature. When its surface has been skimmed of all impurities it does not tarnish. Aluminium is very slow to melt, not only because its specific heat is considerable, but its latent heat appears very large. It is best to make a small fire and then wait patiently till it melts. One can very well work with an uncovered crucible. When it is desired to melt pieces together, they can be united by agitating the crucible or compressing the mass with a well-cleaned, cylindrical bar of iron. Clippings, filings, etc., are melted thus: Separate out first, as far as possible, foreign metals, and to avoid their combining with the aluminium heat the divided metal to as low a heat as possible, just sufficient to melt it. The oil and organic matters will burn, leaving a cinder, which hinders the reunion of the metal if one does not press firmly with the iron bar. The metal may then be cast very easily, and there is found at the bottom of the crucible a little cinder which still contains a quantity of aluminium in globules. These may be easily separated by rubbing in a mortar and then passing through a sieve, which retains the flattened globules.

Kerl & Stohman: To be able to melt well and pour aluminium, the whole quantity of metal which is to be melted at one time must not be put into the crucible at once, but little by little, so increasing the mass from time to time as the contents become fully melted. The necessary knack for attaining a good clean melt consists in dipping the pieces which are to be melted together in benzine before putting in the crucible. Mourey even pours a small quantity of benzine into the crucible after the full melting of the metal, and he recommends the employment of benzine in the melting of all the noble metals. Turning to the cases arising in the employment of aluminium in the different industrial arts, one must as far as possible separate out first the pieces which have been soldered, in order that the newly melted aluminium may not be contaminated by the solder. The solder adhering to these pieces can be removed by treating them

WORKING IN ALUMINIUM.

with nitric acid, by which the aluminium is not attacked.

Mierzinski: To melt aluminium one cannot heat it in common clay crucibles, because it reduces silicon from them, by which the metal becomes gray and brittle. This difficulty can be removed by lining the crucible with carbon, or better, with well-burnt cryolite-clay. Moreover, in practice, it is only in the rarest cases that pure aluminium is obtained to be melted up, but, as a rule, it is alloyed with four to eight per cent. of silver.

CASTING ALUMINIUM.

Deville: Aluminium can be cast very easily in metallic moulds, but better in sand for complicated objects. The mould ought to be very dry, made of a porous sand, and should allow free exit to the air expelled by the metal, which is viscous when melted. The number of vents ought to be very large, and a long, perfectly round git should be provided. The aluminium, heated to redness, ought to be poured rather quickly, letting a little melted metal remain in the git till it is full, to provide for the contraction of the metal as it solidifies. In general, this precaution ought to be taken even when aluminium is cast in iron ingot moulds or moulds of any other metal. The closed ingot moulds give the best metal for rolling or hammering. By following these precautions, castings of great beauty may be obtained, but it is not advisable to conceal the fact that to be able to succeed completely in all these various operations requires for aluminium, as for all other metals, a special familiarity with the material which practice alone is able to give.

In the fusion of impure aluminium, very different phenomena are observed according to the nature of the foreign metal which contaminates it. Ferruginous material often leaves a skeleton less fusible and pretty rich in iron; a liquation has taken place, increasing the purity of the melted material. When the aluminium contains silicon this liquation is no longer possible, or at least it is very difficult, and I have sometimes seen some commercial aluminium so siliceous that the workmen were unable to remelt it. But the aluminium which is made at present is much purer than that.

PURIFICATION OF ALUMINIUM.

Freeing from Slag.—Deville gives the following information on this important subject:—

"It is of great importance not to sell any aluminium except that which is entirely free from the slag with which it was produced and with which its whole mass may become impregnated. We have tried all sorts of ways of attaining this end, so as to obtain a metal which would not give any fluorides or chlorides upon boiling with water, or give a solution which would be precipitated by silver nitrate. At Glacière, we granulated the metal by pouring it while in good fusion into water acidulated with H^2SO^4; this method partially succeeded. But the process which M. Paul Morin uses at present, and which seems to give the best results, is yet simpler. Three or four kilos of aluminium are melted in a plumbago crucible without a lid, and kept a long time red hot in contact with the air. Almost always acid fumes exhale from the surface, indicating the decomposition by air or moisture of the saline matter impregnating the metal. The crucible being withdrawn from the fire, a skimmer is put into the metal. This skimmer is of cast iron; its surface ought not to be rough and it will not be wetted by the aluminium in the least during the skimming. The white and slaggy matters are then removed, carrying away also a little metal, and are put aside to be remelted. So, in this purification, there is really no loss of metal. After having thus been skimmed, the aluminium is cast into ingots. This operation is repeated three or four times until the metal is perfectly clean, which is, however, not easily told by its appearance, for, after the first fusion, the crude aluminium when cast into ingots has a brilliancy and color such as one would judge quite irreproachable, but the metal would not be clean when it was worked, and especially when polished would present a multitude of little points called technically 'piqûres,' which give to its surface, especially with time, a disagreeable look. Aluminium, pure and free from slag, improves in color on using. It is the contrary with the impure metal or with aluminium not freed from slag. When aluminium is submitted to a slow, corroding action, its surface will cover itself uniformly with a white, thin coating of alumina. However, any time that this layer is black or the aluminium tarnishes, we may be sure that it contains a foreign metal and that the alteration is due to this impurity."

Watts suggests that the iron skimmer be oxidized on its surface.

Freeing from Impurities.—Again Deville is the authority, and we quote his advice on the subject:—

"A particular characteristic of the metallurgy of aluminium is that it is necessary, in order to get pure metal, to obtain it so at the first attempt. When it contains silicon, I know of no way to eliminate it, all the experiments which I have made on the subject have had a negative result; simple fusion of the metal in a crucible, permitting the separation by liquation of metals more dense, seems rather to increase the amount of silicon than to decrease it. When the aluminium contains iron or copper, each fusion purifies it up to a certain limit, and if the operation is done at a low heat there is found at the bottom of the crucible a metallic skeleton containing much more iron and copper than the primitive alloy. At first I made this liquation in the muffle of a cupel furnace, in which process the access of air permitted the partial oxidation of these two metals. The little lead which aluminium may sometimes take up may thus be easily separated. Unfortunately, the process does not give completely satisfactory results. It is the same in fusing impure aluminium under a layer of potassium sulphide, K^2S^3; there is a partial separation of the lead, copper, and iron. That which has succeeded best with us is the process which we have employed for a long time at Glacière, and which consists in melting the aluminium under nitre in an iron crucible. We have in this way improved the quality of large quantities of aluminium. The operation is conducted as follows: Aluminium has generally been melted with nitre in order to purify it by means of the strong disengagement of oxygen at a red heat, no doubts being entertained as to the certainty of the result. But it is necessary to take great care when doing this in an earthen crucible. The silica of the crucible is dissolved by the nitre, the glass thus formed is decomposed by the aluminium, and the siliceous aluminium thus formed is, as we know, very oxidizable, and especially in the presence of alkalies. So, the purification of aluminium by nitre ought to be done in a cast-iron crucible well oxidized itself by nitre on the inside.

"On melting aluminium containing zinc in contact with the air and at a temperature which will volatilize the zinc, the largest part of the latter burns and disappears as flaky oxide. To obtain a complete separation of the two metals it is necessary to heat the alloy to a high temperature in a brasqued crucible. This experiment succeeds very well, but it is here shown that the aluminium must oxidize slightly on its surface, for some carbon is reduced by the aluminium from the carbonic oxide with which the crucible is filled. This carbon thus separated is quite amorphous."

This phenomenon may not appear so extraordinary if we consider the case in this way: The aluminium is dissolved in the fluid zinc in a manner strictly analogous to the aluminium dissolved in mercury. Now, it will be seen that aluminium-amalgam decomposes easily, the mercury appearing to impart to the aluminium the ability to combine easily with oxygen, so that in the amalgam aluminium is said to play the part of an alkali metal, with which it is so closely related in its compounds. Considering the case of the aluminium dissolved in melted zinc instead of in mercury, it appears probable that the zinc imparts in the same manner as the mercury, though not necessarily in the same degree, the alkali characteristics to the metal, causing it to oxidize even at the expense of carbonic oxide.

Mierzinski recommends the purification with

nitre to be made in a crucible made of alumina or aluminate of soda.

G. Buchner* states that commercial aluminium contains considerable quantities of silicon, which by treatment, when melted, with hydrogen, evolves hydrogen silicide. This does not result if arsenic is present.

Mallet made chemically pure aluminium by treating the commercial metal with bromine, purifying the resulting Al^2Br^6 by fractional distillation, and then reducing it with pure sodium. By repeatedly melting the metal upon aluminium leaf, he obtained it chemically pure. Although this method is quite applicable when studying the properties of the pure metal, yet it cannot serve on an industrial scale.

Uses of Aluminium.

"Since aluminium was prepared by Deville† on a large scale it has received numerous applications. Its beautiful color, its lightness, its unoxidizability in contact with air or sulphuric acid, its harmlessness to the health, the ease with which it may be worked, are some of the properties which assure for it a place among the useful metals. On account of its very high price the first articles made of it were those of ornament and luxury. The very first article

* Wagner's Jahresb., 1884.
† Fremy's Ency., 1883.

made of it was a baby rattle, intended for the young Prince Imperial in 1856. Afterwards there were made of it jewelry, medals, inlaid work, and carved mouldings for inlaid work and rich furniture. It is very well suited for fine jewelry by reason of its adaptability to being cast and carved, the beautiful reflections from a chased surface, its color, which matches well with gold, and its absence of all odor. Later on, the lightness of aluminium leads to its use for telescope tubes, marine glasses, eye-glasses, and especially sextants. In delicate physical apparatus, where it is necessary to avoid the inertia of large masses, aluminium replaces the other metals with advantage. It is used for beams for delicate balances and for very small weights. There have been made of it sabre sheaths, sword handles, and the imperial eagles for the French army. Finally, made into fine wire, it is worked into lace, embroidery, etc. For all these purposes aluminium answers better than silver, for the objects are much lighter and do not tarnish. The resistance of aluminium to most of the agents which attack the useful metals has led to its employment for culinary articles; a large number of which were seen at the London Exhibition in 1862. But the advantages of aluminium vessels have not yet been sufficiently comprehended, and this use of it has at present been entirely discontinued. Likewise, aluminium jewelry is not seen any more; so that the metal seems reserved for little more than optical and surgical instruments. But the aluminium industry is nevertheless established on a permanent basis and will continue, because of the numerous applications of its alloys."

M. Dumas made a helmet of aluminium, gilded and ornamented, which weighed complete only one and one-fifth pounds.

Aluminium leaf, beaten very thin, may be used anywhere in place of silver leaf. It is applied in the same manner, and is more durable.

Aluminium wire has been proposed for telegraph lines. The conductivity of aluminium is double that of iron, and as it is so much lighter, thinner wire can be used. As its high price is a practical difficulty, an alloy of iron and aluminium has been suggested.

"One of the most likely applications of aluminium is probably as a material for statuettes and small works of art of this description, especially if the means could be found of giving to it a richer color and appearance either by a kind of bronzing or some alloy.

"Aluminium makes very bright reflectors, not tarnished by the products of combustion, while the slight bluish tinge of the metal corrects the yellowish tinge of the flame. For culinary uses it is well adapted, because of its lightness and the little tendency it has to become corroded by any of the liquids likely to come in contact with it. It is necessary to observe, however, that this power of resisting the action of corroding agencies, and more especially the atmosphere of large towns, is exhibited only by the pure metal. Most of the metal of commerce is very impure with iron and silicon, not having been properly freed from slag. Aluminium thus contaminated soon becomes tarnished, and much disappointment has been experienced from this cause by those who have used it for ornamental purposes. According to Deville, the impurities just mentioned are found to the greatest amount in the metal obtained from cryolite."

In the 'Scientific American,' vol. xii. pp. 31 and 51, is a long article on plating with aluminium, giving complete directions for preparing articles, solutions, etc.

A large collection of articles of aluminium was shipped from England to Calcutta in Oct. 1883, intended for exhibition there. The exhibit consisted of wire, pens, pencil-cases, railway-carriage fittings, locks and bolts, harness furniture in great variety, chandeliers, cutlery, and ships' fittings, and illustrated very well the various uses to which the metal can be put. It is being used for the lighter parts of such instruments as galvanometers, etc., for suture wire, and perhaps its most promising field is for engineering, astronomical, and optical instruments.

" Aluminium is sold as leaf in books, like gold leaf, for decorations, at from 40 to 50 cents per book, and is being experimented with by manufacturers of jewelry. In Germany, experiments have been made with it as a coating for iron, to be applied for ornamental purposes, and as an improvement on tin plate. Its use is extending slowly but surely, its cost being at present the principal obstacle to its wider employment."*

Experiments were made in the U. S. Mint in 1865, on alloys of aluminium for coins. The results were not sufficiently successful to induce the Government to adopt the metal for that purpose.

Soldering Aluminium.

At the time Deville wrote his book, the difficulty of soldering aluminium properly was one of the greatest, if not the greatest, obstacle to the employment of the metal. His views on the question may be, therefore, very interesting; they are as follows:—

"Aluminium may be soldered, but in a very imperfect manner, either by means of zinc or cadmium, or alloys of aluminium with these metals. But a very peculiar difficulty arises here, we know no flux to clean the aluminium which does not attack the solder, or which, protecting the solder, does not attack the aluminium. There is also an obstacle in the particular resistance of aluminium to being wetted by the more fusible metals, and on this account the solder does not run between

* Mineral Resources of the U. S. 1883–4.

and attach itself to the surfaces to be united. M. Christofle and M. Charrière made, in 1855, during the Exposition, solderings with zinc or tin. But this is a weak solder and does not make a firm seam. MM. Tissier, after some experiments made in my laboratory, proposed alloys of aluminium and zinc, which did not succeed any better. However, M. Denis, of Nancy, has remarked that whenever the aluminium and the solder melted on its surface are touched by a piece of zinc, the adhesion becomes manifest very rapidly, as if a particular electrical state was determined at the moment of contact. But even this produces only weak solderings, insufficient in most cases.

"A long time ago, M. Hulot proposed to avoid the difficulty by previously covering the piece with copper, then soldering the copper surfaces. To effect this, plunge the article, or at least the part to be soldered, into a bath of acid sulphate of copper. Put the positive pole of a battery in communication with the bath, and with the negative pole touch the places to be covered, and the copper is deposited very regularly. M. Mourey has succeeded in soldering aluminium by processes yet unknown to me; samples which I have seen looked excellent. I hope, then, that this problem has found, thanks to his ingenuity, a solution; a very important step in enlarging the employment of aluminium."

Mierzinski gives the following statements about M. Mourey's solder:—

"Mourey, who first made a practicable solder for aluminium, used two kinds of solder, soft and hard. The first was used for the usual soldering up of flasks or pieces of metal. He made solders of five different alloys, the composition of which were as given in the table below:—

	I.	II.	III.	IV.	V.
Al	20	15	12	8	6
Zn	80	85	88	92	94

These solders have varying melting points, and thus there results the hard and soft solders. One can take a soft solder, as IV., for brazing, and one like II. for ordinary soldering."*

Schwarz† improved these solders by adding copper to the alloy. His solders have the following composition:—

	I.	II.	III.	IV.	V.
Al	12	9	7	6	4
Cu	8	6	5	4	2
Zn	80	85	88	90	94

Mourey‡ recommends improved solders of somewhat similar composition. They are:—

	I.	II.	III.	IV.	V.	VI.	VII.
Al	30	20	12	9	7	6	4
Cu	20	15	8	...	...	...	2
Brass	...	...	...	6	5	4	...
Zn	50	65	80	85	88	90	94

---

* It is usual to employ hard solder for brazing, and No. II. would be harder than No. IV.—J. W. R.
† Dingler, 157, 445.   ‡ Dingler, 166, 205.

Col. Wm. Frishmuth* recommends a solder containing:—

Al	20
Cu	10
Zn	30
Sn	60
Ag	10

Col. Frishmuth† states that the solder just given is used for fine ornamental work, while for lower-grade work he uses the following:—

	I.	II.	III.
Sn	95	97	98–99
Bi	5	3	2–1

Frishmuth recommends for a flux, in all cases, either paraffin, stearin, vaselin, copaiva balsam, or benzine. In the solder for fine work, if aluminium is used in larger quantity than recommended, the solder becomes brittle.

Kerl and Stohman give the following practical observations on this subject:—

"At first, the soldering of aluminium appeared impossible. But Ph. Mourey, a gold and silver worker in Paris, invented a new method by which he could solder any kind of object of this metal. The following are his receipts:—

"There are needed, according to the objects to be soldered, five different solders, which are composed of aluminium, copper, and zinc, in different proportions:—

---

* Techniker, vi. 249.   † Wagner's Jahresb., 1884.

	I.	II.	III.	IV.	V.
Al.	12	9	7	6	4
Cu.	8	6	5	4	2
Zn.	80	85	88	90	94

"To make the solder, first put the copper in the crucible. When it is melted, then add the aluminium in three or four portions, thereby somewhat cooling the melted mass. When both metals are melted, the mass is stirred with a small iron rod, and then the required quantity of zinc added, free from iron, and as clean as possible. It melts very rapidly. The alloy is then stirred briskly with an iron rod for a time, some fat or benzine being meanwhile put in the crucible to prevent contact of the metal with air and oxidation of the zinc. Finally the whole is poured out into an ingot mould previously rubbed with benzine. After the addition of zinc, the operation must be finished very rapidly, because the latter will volatilize and burn out. As soon as the zinc is melted, the crucible is taken out of the fire.

"The separate pieces of metal to be soldered together are first well cleaned, then made somewhat rough with a file at the place of juncture, and the appropriate solder put on it in pieces about the size of millet grains. The objects are laid on some hot charcoal, and the melting of the solder effected by a blast lamp or a Rochemont turpentine-oil lamp. During the melting of the solder, it is rubbed with a little soldering iron of pure aluminium. The soldering iron of pure aluminium is essentially a necessity for the success of the operation, since an iron of any other metal will alloy with the metals composing the solder, while the melted solder does not stick to the iron made of aluminium.

"The method just described differs from the one described by Mourey in so far that he used, at first, alloys of aluminium and zinc only, with no copper. He used one of the more fusible alloys to first unite the pieces, and then used a less fusible one to finish with. In order to avoid the oxidation of the solder he added while using the hard solder, which must be worked with a hotter iron, a quantity of copaiva balsam and turpentine, which acts just as borax in working silver. With these new solders of aluminium, copper, and zinc the process is much simpler, the work is done with one solder and the moistening with balsam is unnecessary. The solderings may be done so perfectly that plates soldered together never break at the joint when bent back and forth, but always give way in other places; which is a result not always possible in the best soldering of plates of silver."

Bell Bros. used to operate the works at Newcastle-on-Tyne, and their description may contain a few points not yet brought forward :—*

"In order to unite pieces of aluminium, small

---

* Chem. News, iv. 81.

tools of the same metal are used, which facilitate at the same time the fusion of the solder and its adhesion to the previously prepared surfaces. Tools of copper or brass must be strictly avoided, as they would form colored alloys with the aluminium and the solder. The use of the little tools of aluminium is an art which the workman must acquire by practice. At the moment of fusion the work needs the application of friction, as the solder suddenly melts very completely. In soldering it is well to have both hands free and to use only the foot for the blowing apparatus. The solders used are of aluminium, copper, and zinc. (See the ones given by Kerl & Stohman, p. 250.) No. IV. is the one generally preferred, particularly for small objects. In order to make the solder, the copper is first melted, the aluminium added, and the whole stirred with an unpolished iron rod, just as it comes from the forge, adding also a little tallow. The zinc is then added, avoiding too much heat, which would drive it off. In soldering, also, too high a heat should be avoided for the same reason.

Veneering with Aluminium.

Deville is the first writer to make mention of this art:—

"M. Sevrard succeeded in 1854 in plating aluminium on copper and brass with great perfection. The two metallic surfaces being prepared in the ordinary manner and well scoured with sand, they are placed one on the other and held tightly between two iron plates. The packet is then heated to dark redness, at which temperature it is strongly compressed. The veneer becomes very firmly attached, and sheets of it may be beaten out. I have a specimen of such work perfectly preserved. The delicate point of the operation is to just heat the packet to that point that the adherence may be produced without fusing the aluminium, for when it is not heated quite near to this fusing point, the adherence is incomplete. Experiments of this kind with copper and aluminium foil did not succeed, for as soon as any adherence manifested itself the two metals combined and the foil disappeared into the copper. In an operation made at too low a temperature, the two metals, as they do not behave similarly on rolling, become detached after a few passes through the rolls. Since then, the experiments in veneering aluminium on copper, with or without the intervention of silver, have succeeded very well."

The only other article to be found on this subject is Dr. Clemens Winckler's paper, from which we extract the following :—*

"The question demands attention whether it is not possible to coat certain metals and alloys with aluminium, and thereby impart to them, superfi-

* Industrie Blätter, 1873.

WORKING IN ALUMINIUM. 255

cially at least, the advantageous properties of that metal. The present high price of the metal does not stand in its way for this purpose; and it only remains now to decide whether it is practicable to coat our common metals, iron, copper, etc., with it. The question must at present be answered in the negative. Two methods can be used for covering one metal with another, galvanoplasty and plating or veneering. The separation of aluminium by the galvanic current succeeds only by the use of a bath of molten anhydrous $Al^2Cl^6.2NaCl$, melting at 165° C. (329° F.), but the metal is deposited as a non-coherent powder, mixed with NaCl, and therefore the object of plating is not attained in this way. No one has yet been able to throw down aluminium in a metallic state from aqueous solution, and it was an error when Gore stated that he had coated copper with aluminium by means of a solution of Al^2Cl^6 in water and a weak galvanic current. Concerning the coating of metals by the so-called plating method, it is indeed, according to my own experience, possible to a certain degree, but the product is entirely useless, every plating requiring an incipient fusing of both metals and their final intimate union by rolling. The ductility of aluminium is, however, greatly injured by even a slight admixture with other metals; iron makes it brittle and copper, in small per cent. makes it fragile as glass. If now it were possible in any way to fuse a coating of aluminium upon another metal, there would be formed an intermediate alloy between the two metals from which all ductility would be gone and which would crumble to powder under the pressure of the rolls, thus separating the aluminium surface from the metal beneath. But even if it were possible in this way to coat a metal with a thin plate, it is still doubtful if anything would be attained thereby. For, while compact aluminium resists oxidizing and sulphurizing agencies, the divided metal does not. In powder or leaves aluminium is readily oxidized, as is shown by its amalgam becoming heated in the air and quickly forming alumina. In the form of a coating upon other metals it must necessarily be in a somewhat finely divided state, and hence would probably lose its durability."

Gilding and Silvering Aluminium.

Deville says: "The gilding and silvering of aluminium by electricity is very difficult to do satisfactorily and obtain the desirable solidity. M. Paul Morin and I have often tried it by using a bath of acid sulphide of gold or of nitrate of silver with an excess of sulphurous acid. Our success has only been partial. However, M. Mourey, who has already rendered great services in galvanoplasty, gilds and silvers the aluminium of commerce with a surprising perfection considering the little time he has had to study the question. I also know that Mr. Christofle has gilded it, but I am entirely ignorant of the methods employed by these gentlemen."

Watts's Dictionary: "Eight grammes of gold are dissolved in aqua regia, the solution diluted with water and left to digest twenty-four hours with an excess of lime. The precipitate, with the lime, is well washed, and then treated with a solution of twenty grammes of hyposulphate of soda. The liquid resulting serves for the gilding of aluminium without the aid of heat or electricity, the metal being simply immersed in it after being previously well cleaned by the successive use of caustic potash, nitric acid, and pure water."

Kerl and Stohman: "Gilding and silvering aluminium galvanically does not offer the least difficulty. One can, by using a proper ground, coat it with silver and gold in six different colors, by employing the correct combination, such as shining or matt gold and silver or lead gray."

PART XII.

ALLOYS OF ALUMINIUM.

General Remark.—Mierzinski:—

"Aluminium unites easily with most metals, the combination being usually accompanied by a lively disengagement of heat. Quite homogeneous alloys can be made, which for the most part are easily worked and have important applications. The alloys in general become harder the greater the proportion of aluminium, and become brittle if this proportion passes a certain limit, which with gold and copper is very low. On addition of a larger amount of aluminium than this limit allows, gold and copper become whiter, and at last entirely lose their color. The addition of other metals to aluminium imparts to it the same new properties. It becomes brighter and somewhat harder, but, united with small quantities of zinc, tin, gold, or silver, remains malleable. Iron and copper impart to it no specially prejudicial qualities, if they are not present in too large quantities. The alloys most frequently used are those of copper, silver, and tin. These owe their numerous uses to their fine color, their resistance to most chemical agents, and the facility with which they may be worked."

Aluminium and Silicon.

Tissier: "As Deville has observed, silicon is far from injurious to the malleability of aluminium, the latter bearing it much as iron and copper do. We have had occasion to analyze a specimen of aluminium, which, although it worked with difficulty, was yet employed to make various objects, and yet, attacked by HCl, it left an insoluble residue of no less than 15.67 per cent. But, even admitting that this residue still retained some aluminium with the silicon, we think that there was at least 10 per cent. of the latter in this specimen."

Deville: "Any siliceous material whatever, put in contact with aluminium at a high temperature, is always decomposed; and if the metal is in excess there is formed an alloy or a combination of silicon and aluminium in which the two bodies may be united in almost any proportions. Glass, clay, and the earth of crucibles act in this way. However, aluminium may be melted in glassware or earthen crucibles without the least contamination of the metal if there is no contact between the metal and the material; the aluminium will not wet the crucible if put into it alone. But, the moment that any flux whatever facilitates immediate contact, even sodium chloride does this, the reaction begins to take place, and the metal obtained is always more or less siliceous. It is for this reason that I have prescribed in melting aluminium not to add any kind of flux, even when the flux would not be attacked by the metal. Among the fusible materials which facilitate the melting of aluminium, it is necessary to remark of the fluorides that they attack the siliceous materials of the crucible, dissolving them with great energy, and then the siliceous materials thus brought into solution are decomposed by the aluminium with quite remarkable facility. Aluminium charged with silicon presents quite different qualities according to the proportion of the alloy. When the aluminium is in large excess, there is obtained what I have called the 'cast-iron' state of aluminium, by means of which I discovered crystallized silicon in 1854. This 'cast' aluminium, gray and brittle, contains according to my analysis 10.3 per cent. of silicon and traces of iron. When siliceous aluminium is attacked by hydrochloric acid, the hydrogen which it disengages has an infected odor, which I formerly attributed to the presence of a hydrocarbon, but which we now know is due to hydrogen silicide, SiH^4, thanks to the fine experiments of MM. Wöhler and Buff. It is by the production of this gas that may be explained the iron smell which is given out by aluminium more or less contaminated with silicon. But aluminium may absorb much larger proportions of silicon, for, on treating fluo-silicate of potash with aluminium, M. Wöhler obtained a material still metallic containing about 70 per cent. of silicon, sometimes occurring as easily separable crystals. Since I had the occasion in a work which I published on silicon to examine a large number of these combinations, I found that they were much more alterable than pure aluminium or silicon, without doubt because of the affinity which exists between silica and alumina. I have, therefore, dwelt on and tried to explain the importance of this point in obtaining perfectly pure aluminium. I should say, in addition, that the metal now sold in commerce may contain either iron or silicon, according to the method of preparation. These two impurities are hurtful to most of the qualities of the aluminium, and everything ought to be done to avoid their presence."

ALUMINIUM AND MERCURY.

Deville: "Mercury is not able to unite with aluminium. Experiments of this nature which I have made myself, and which Mr. Wollaston has confirmed, prove it most clearly."

Watts: "According to Caillet, aluminium may be amalgamated by the action of ammonium or sodium amalgam, with water; also when it is connected with the negative pole of a voltaic battery and dipped into the mercury moistened with acidulated water, or into a solution of mercuric nitrate. Tissier* confirms this statement respecting the

* Compt. Rend., xlix. 56.

battery method, and adds that if the aluminium foil is not very thick it becomes amalgamated throughout and very brittle." Tissier also finds that aluminium may be made to unite with mercury merely by the intervention of a solution of caustic potash or soda, without the intervention of the battery. If the surface of the metal be well cleaned, or moistened with the alkaline solution, it is immediately melted by the mercury, and a shining amalgam forms on its surface. The amalgam of aluminium instantly loses its lustre when exposed to the air, becoming heated and rapidly converted into alumina and mercury. It decomposes water with evolution of hydrogen and formation of alumina and mercury. Nitric acid attacks it with violence.

Watts (First Supplement) states that aluminium amalgam may be formed either by bringing the aluminium in contact with mercury containing a small quantity of sodium, or by Joules's method of electrolyzing the solution of an aluminium salt, with mercury for the negative pole;* but the best method is to heat the two metals together in a gas which does not act on either of them. To do this, a piece of aluminium foil is placed at the bottom of a thick-walled test-tube, and well-dried mercury is poured on it, the tube having been previously drawn out at the middle to prevent the foil rising

---

* Chem. Gazette, 1850, p. 339.

to the surface. The air is then expelled by a stream of carbonic acid gas and the tube is heated, without interrupting the current of gas, till the metal is all dissolved.

Aluminium amalgam decomposes in contact with air or water more quickly than sodium amalgam. When a few drops of an amalgam containing but a small proportion of aluminium are left in contact with moist air, gelatinous, opalescent excrescences of pure hydrated alumina are seen to form on their surfaces, exhibiting both in their form and mode of growth considerable resemblance to the so-called Pharoah's serpents. This hydrated alumina is perfectly soluble in acids and alkalies. Water has the same effect as moist air. Watts, in vol. viii., states that aluminium oxidizes when its surface is rubbed with a piece of soft leather impregnated with mercury. The rubbed surface becomes warm, and in a few seconds whitish excrescences appear, consisting of pure alumina. The presence of mercury appears necessary to produce the result.

Fremy says that Tissier has proven that aluminium previously contaminated with caustic potash or soda combines easily with mercury. The alloy which results is very brittle, the aluminium in it decomposes water, oxidizes easily in the air, and behaves as a metal of the alkaline earths.

Gmelin* states that potassium amalgam intro-

* Hand Book, vi. 3.

duced into a hole bored in a crystal of alum immediately acquires a rotary motion, which lasts sometimes half an hour. At the same time, it takes up a considerable quantity of aluminium and becomes more viscid.

Aluminium and Copper.

Tissier Bros., 1858: "Just as copper increases the hardness of aluminium, so aluminium in small proportions increases the hardness of copper. However, aluminium does not injure its malleability, but makes it susceptible of taking a beautiful polish, and, according to the proportions, varies its color from red gold to pale yellow. These facts were announced some time back by Dr. Percy, in England, who made the alloy by introducing copper into the mixture of cryolite and sodium which he was reducing. We have made large quantities of these alloys, and we may say that they leave nothing to be desired in regard to lustre or color to make them perfect imitations of gold. They alter much less by successive fusions than the alloys of copper with zinc and tin employed for the same object. A ten per cent aluminium alloy was harder than our gold coin, took a fine polish by burnishing, and had the color of pale jeweller's gold; it could be forged and worked the same as copper. The five per cent. aluminium alloy was less hard than the preceding, but, like it, takes a fine polish, and in tint approaches nearly to pure gold. The twenty per cent. aluminium alloy much resembles bismuth, having a whitish-yellow tint. This alloy crystallizes in large leaves and pulverizes in the mortar like bismuth or antimony. Alloys with five to ten per cent. of aluminium may have their color changed at will, either by leaving in nitric acid, which takes away the copper and leaves the aluminium, or in hydrochloric, which leaves the copper. The resistance, hardness, and elasticity, which are communicated to copper by introducing small quantities of aluminium, will certainly make these important industrial alloys."

Deville, 1859: "The aluminium and copper alloys with two to three per cent. aluminium are used by M. Christofle, who employs them for large castings of objects of art. They are harder than aluminium, and work well under the burin and chisel. The alloy with ten per cent. aluminium had its useful properties first described by M. Debray. It is very hard, can be beaten out cold, but with remarkable perfection when hot, and may be well compared to iron, which it resembles in all these physical properties. It is also very ductile. This ten per cent. aluminium alloy is usually known as aluminium bronze. It behaves as a true alloy, and, in consequence, will not liquate into different combinations. It is formed of—

9 equivalents of Cu . . . 275   9
1   "   " Al . . .   28   1
                                          303   10

This is proven by the fact that, when in making the alloy the pure copper is in the crucible and a bar of aluminium is added, the combination takes place with such disengagement of heat that if the crucible is not of good quality it will be fused, for the whole becomes white hot.

"The color of the ingot of bronze is exactly that of 'green gold,' an alloy of gold and silver. The bronze receives a beautiful polish, being comparable in this regard only to steel. Its chemical properties do not differ much from those of most of the alloys of copper. However, in numerous experiments, we have noticed that it resists most chemical agents much better than these, especially sea-water and sulphuretted hydrogen. Its tenacity is equal to that of steel. M. Lechatelier made the following determinations on the metal cast into cylinders:

Per cent. of aluminium.	Diam. of cylinder.	Breaking strain.	Strength per sq. m. m.
10	10.0 m. m.	4027 kilos.	58.36 kilos.
10	10.1 "	4432 "	55.35 "
8	10.1 "	2657 "	33.18 "
5	10.1 "	2582 "	32.20 "
5	10.1 "	2517 "	31.43 "
	French wrought iron	. .	35.00 "

"A. Gordon made some experiments recently, in which the strength of the aluminium bronze which he tested was 84.00 kilos per square millimetre. I made the test on some wire, and the result I reached was 85.00 kilos; under the same conditions iron gave 60.00 and best steel 90.00 kilos. According to experiments as to its wear as journal boxes, it is found to wear away less than any other journal metal yet tried.

"Its malleability is almost perfect, as is seen by the following report of M. Boudaret, a practical engineer: First, aluminium bronze is malleable at all temperatures, from bright red to cold; second, it is perfectly malleable at red heat, breaking less and elongating more than pure copper; third, it is hard to roll in the cold, after several passes it ceases to elongate and must then be annealed very often or it will break quickly; fourth, it results from the foregoing that it is best to roll it at as high a heat as possible below fusion; fifth, annealing and tempering render it softer than simple annealing. If after having annealed at bright red heat it is let cool in still air to redness and then plunged into cold water, it is ductile and malleable enough in the cold to stand all industrial working."

Mierzinski, 1885: "Two points are to be attended to in making aluminium bronze. First, a very pure copper must be used, the best is that electrically deposited, but it generally costs too much. The next best is the Lake Superior brand. The usual commercial copper gives all sorts of poor results, owing to the antimony, arsenic, tin, zinc, or iron contaminating it. The bronze loses by being alloyed with zinc or tin. Second, the alloy must be remelted two or three times to remove its brittleness. In all probability, the percentage of aluminium increases by remelting. The usual alloys are those with 1, 2, 5, and 10 per cent. aluminium. The 5 per cent bronze is golden in color, polishes well, casts beautifully, is very malleable cold or hot, and has great strength, especially after hammering; its defect is that it easily oxidizes or tarnishes. The 7.5 per cent. bronze is to be recommended as superior to the 5 per cent.; it has a peculiar greenish-gold color, which makes it very suitable for decoration. All these good qualities are possessed by the 10 per cent. bronze. It is bright golden, keeps its polish in the air, may be easily engraved, shows an elasticity much greater than steel, and can be soldered with hard solder. It gives good castings of all sizes and runs in sand moulds very uniformly. Thin castings come out very sharp, but if a casting is thin and suddenly thickens, small offshoots must be made at the thick place into which the metal can run and then soak back into the casting as it cools and shrinks, thus avoiding cavities by shrinkage at the thick part. Its sp. gr. is 7.689, that of soft iron. Its strength, when cast, is between that of iron and steel; but when hammered it is equal to best steel. It may be forged at about the same heat as cast steel, and then hammered until it is almost cold without breaking or ripping. Tempering makes it soft and malleable. It does not foul a file, and may be easily drawn into wire. Any part of a machine which is usually made of steel can be replaced by this bronze. As a solder for it, Hulot uses an alloy of the usual half-and-half lead-tin solder with 12.5, 25, or 50 per cent of zinc amalgam."

Fremy: "By the addition of a small amount of copper, aluminium becomes hard, brittle, and takes a bluish-white color. The alloy with 5 per cent. aluminium is very malleable, but if over 10 per cent. Al is present the alloy cannot be used. The 10 per cent. bronze is now replacing ordinary bronze in the manufacture of articles which are to stand great resistance, such as axle bearings, weavers' shuttles, etc. Reflectors are also made of it, for the smoke of oil, like illuminating gas, does not tarnish it. By whatever method these bronzes are made, they are at first very brittle, but by a series of successive fusions and solidifications they may be made to acquire the necessary solidity and tenacity."

Kerl and Stolman: "Most of the copper-aluminium alloys are very brittle and easily oxidized. Only the 5 to 10 per cent. aluminium alloys are fixed, forgeable, tenacious, and of fine color. Alloys with much aluminium and little copper are not forgeable, and are bluish or grayish-white. With 60 to 70 per cent. aluminium they are very brittle, glass hard, and beautifully crystalline.

With 50 per cent. the alloy is quite soft, but under 30 per cent. of aluminium the hardness returns."

'Chemical News,' vii. p. 220, contains a long paper on testing aluminium bronze (10 per cent.) as to its suitability for the construction of astronomical and philosophical instruments, the work of an English Royal Engineer. He concludes his observations with these words: "It appears from these experiments that the 10 per cent. bronze is far superior, not in one or in some but in every respect, to any metal hitherto used for these instruments. Its sp. gr. is 7.689, strength 73,185 pounds per square inch, to that of gun metal 35,000; it is malleable almost to its melting point, and can be soldered with either brass or silver solder."

'Chemical News,' v. p. 138, contains a number of experiments on the relative strengths of these alloys. The results are as follows, the numbers expressing the results being merely relative:—

	Cu	Sn	Al	Strength.
Ordinary gun metal, 11 per cent tin and 89 per cent. copper				10
Copper, with 10 per cent. aluminium				19
Drawn copper-wire				7
Drawn brass-wire				8
Tertiary alloys	96	4	0	4
	96	4	1	10
	96	4	2	16

Bell Bros., Newcastle, give the specific gravity of the aluminium bronzes as being—

3 per cent. aluminium . . . 8.691
4 " " . . . 8.621
5 " " . . . 8.369
10 " " , . . . 7.689

'Wagner's Jahresb.' vol. x., contains a long article on aluminium bronze, ten per cent., most of the facts in which have been already given. We may note that the melting-point of this alloy is there stated as about 650°.

Bernard S. Procter,* after describing thirty-one experiments comparing aluminium bronze and brass, sums up the conclusions as follows:—

"From the above experiments it appears that aluminium bronze has a little advantage over ordinary brass in power to withstand corrosion, and its surface, when tarnished, is more easily cleaned. This should give it general preference where cost of material is not an important consideration, especially if strength, lightness, and durability are at the same time desirable. It is out of my power to say anything about its fitness for delicate machinery, except that its chemical examination has revealed nothing which can detract from the preference its mechanical superiority should give it. Being so much less acted on by ammonia and coal-gas suggests its suitability for chemical scales, weights, scoops, etc. Its resistance to the action of the weather and the ease with which tarnish is removed render it especially applicable for door-

* Chem. News, 1861, vol. iv. p. 59.

plates, bell-handles, etc. Its mechanical strength and chemical inactivity together recommend it for hinges exposed to the weather. In experiments 18, 22, etc., the tendency of brass to corrode on the edges and at any roughness on its surface will be observed, while the bronze is free from this defect. In several cases the bronze seemed to be more quickly covered with a slight tarnish which did not increase perceptibly, probably the tarnish acting as a protection to the metal; but the brass, though less rapidly discolored, continued to be corroded and apparently with increased speed as the action was continued. The bronze is more easily cleaned. For culinary vessels its superiority to metals now in use appears questionable. Various philosophical instruments are among the purposes for which the use of the bronze appears advantageous. Undoubtedly, the great obstacle to its extensive application is its high price, resulting partly from the difficulty of getting sufficiently pure copper, the presence of a small amount of iron being very prejudicial." The author states that he wrote the article with a home-made pen of aluminium bronze, and suggests that it is well worthy of the attention of pen-makers.

Thurston[*] says: "The ten per cent. bronze has a tenacity of about 100,000 pounds, compressive strength 130,000 pounds, and its ductility and

---

[*] Materials for Engineering.

toughness are such that it does not even crack when distorted by this load. It is so ductile and malleable that it can be drawn down under a hammer to the fineness of a cambric needle. It works well, casts well, holds a fine surface under the tool, and when exposed to the weather it is in every respect considered the best bronze yet known. Its high cost alone has prevented its extensive use in the arts. The alloys are very uniform in character. Even one per cent. of aluminium added to copper causes a considerable increase in ductility, increases its fusibility, and enables it to cast well; two per cent. gives a mixture used for castings which are to be worked with a chisel. It is softened by sudden cooling from a red heat. Its coefficient of expansion is small at ordinary temperatures. It has great elasticity when made into springs."

Guettier makes the following remarks:—*

"Mr. Strange's experiments in regard to the relative rigidity of brass, ordinary bronze, and aluminium bronze showed that the latter was about forty times as rigid as soft brass and three times as rigid as ordinary bronze. Under the tool, aluminium bronze produces long and resisting chips, and although not entirely unoxidizable, it is not so easily tarnished by air as brass, bronze, or steel."

Knight:† "Aluminium bronze is more difficult

---

* Metallic Alloys, by Guettier.
† American Mechanical Dictionary.

to cut than brass, but cuts very smooth and clean. If less costly it would replace red and yellow brass. In contact with fatty matters or juice of fruit, no soluble metallic salt is formed, which highly recommends it for various articles of table use."

Cowles Bros.* thus describe the alloys of aluminium and copper which they make:—

" In England the Aluminium Crown Metal Co. has for the past three or four years been turning out large quantities of aluminium alloys based on the price of $14.60 per pound for the aluminium in them. Even at the high prices charged, these Webster alloys have attained a great popularity, and are replacing German silver, brass, bronze, etc. Aluminium added to any of the common alloys, such as brass, German silver, or Britannia metal, adds greatly to all their desirable qualities. Aluminium bronze cannot only be used in all places where brass or bronze are now used, but it will likewise soon supersede iron and steel in many places; as for artillery. The maximum standard of strength demanded by the British and German governments in their wrought-steel guns, which cost from 50 cents to $1 per pound, is at present 70,000 pounds tensile strength and 15 per cent. elongation. These guns could be cast of aluminium bronze, giving a greater strength and elongation, at far less cost, being made in one-quarter of the

---

* Cowles' Pamphlet, April, 1886.

time and with a comparatively inexpensive plant. The melting-point of the bronze is somewhat below that of copper and its specific gravity is 7.23. It is without rival as an anti-friction metal, besides having the hardness, tenacity, and wearing qualities of the best steel. It has also the peculiar unctuousness of copper and lead, being so strong and tough that very small quantities of the rolled bronze may be used to bush boxes of cast or wrought iron, so that its first cost is less than that of the thick masses of brass or phosphor-bronze now used. The five per cent. bronze makes beautiful wearing plumbers' goods, and can be used also for table articles, being free from the offensive smell and taste peculiar to brass. Aluminium in almost all proportions up to eight per cent. improves all brasses. Some it makes more ductile, in others it improves the color, and all are greatly increased in strength and power to resist corrosion. The alloy copper 67, zinc 26, aluminium 7 has a strength of 96,000 pounds, while that of copper 67, zinc 30, aluminium 3 has a strength of 65,000 pounds with 12 per cent. elongation. When we understand that ordinary brass rarely has a tensile strength over 30,000 pounds, the extraordinary value of the aluminium can be appreciated. The strength of these alloys on the testing machine is as follows :—

Alloy, Al brass castings.			Tensile strength per sq. in., pounds.	Elongation, per cent.
Al	Cu	Zn		
5.8	67.4	26.8	95,712	1
3.3	63.3	33.3	85,867	7.6
3.0	67.0	30.0	67,341	12.5
1.5	77.5	21.0	32,356	41.7
1.5	71.0	27.5	41,952	27.0
1.25	70.0	28.0	35,059	25.0
2.5	70.0	27.5	40,982	28.0
1.0	57.0	42.0	68,218	2.0
1.15	55.8	43.0	69,520	4.0
Average commercial cast brass			23,000	Less than 10.0

"The second alloy is made by mixing two parts five per cent. aluminium bronze with one part zinc.

"The aluminium bronzes gave the following results:—

Al	Cu	Tensile strength, pounds.	Elongation in per cent.
2.5	97.5	42,770	53
5.0	95.0	68,480	7.8
6.6	93.4	55,638	30
7.5	92.5	54,636	16
7.5	92.5	60,520	22
91	9	87,783	5
90	10	108,966	...
90	10	99,931	1.5
90	10	97,103	3.0
90	10	105,336	7.8
90	10	110,657	5.4
16.8	83.2	29,369	(Sp. gr. 3.23)

"The two 10 per cent. bronzes last given were plunged while red hot into water. Cowles Bros.

are now selling 10 per cent. bronze at forty cents per pound."

In regard to some alloys of aluminium and copper in which other metals are present, we would notice the following alloys which have been made in addition to those already incidentally mentioned.

Aluminium can be melted with brass, argentan, etc., by which new bronzes are made of beautiful color, great hardness, and polish, unalterable in the air, easily cast, etc. One per cent. of aluminium is sufficient to modify the qualities of brass or tin bronze, while 2 per cent. shows a decided change. By taking ordinary bronze with 1 to 2 per cent. of zinc or tin, and adding 1 to 2 per cent. of aluminium, alloys are obtained possessing additional qualities to those of aluminium bronze, and which can replace it in places and for purposes where the latter's qualities are not so well suited.

Besides these simple alloys we have those of copper with nickel, tin, zinc, bismuth, and aluminium, in such quantities as to make any desired color or degree of hardness. The following has a beautiful white polish, which is a close imitation of silver :—

Copper 100
Nickel 23
Aluminium 7

F. II. Sauvage makes a metal resembling pure silver, which he calls Neogen. It contains—

Cu	58
Zn	27
Ni	12
Sn	2
Al	½
Bi	½

"Minargent" is a similar alloy, containing—

Cu	100
Ni	70
Sb	5
Al	2

To make this last alloy, the directions are first to melt together the copper, nickel, and antimony, and then granulate the resulting alloy in water. The dried granules are mixed with the aluminium and with 1.5 per cent. of a flux consisting of 2 parts borax and 1 part fluorspar, and then remelted.

P. Baudrin makes an alloy very much resembling silver in color, malleability, ring, and even sp. gr., of the following composition :—

Cu	75
Ni	16
Zn	2.25
Sn	2.75
Co	2
Fe	1.5
Al	0.5

Jas. Webster[*] patents the following bronze: copper is melted, and aluminium added so as to

---

[*] German Pat., 11,577.

ALLOYS OF ALUMINIUM. 279

make a 10 per cent. bronze, which is then mixed with 1 to 6 per cent. of an alloy of—

Cu	20
Ni	20
Sn	30
Al	7

Thos. Shaw, of Newark, N. J.,* patents a phosphor aluminium bronze, making the following claims: First, an alloy of copper, aluminium, and phosphorus containing 0.33 to 5 per cent. of aluminium, 0.05 to 1 per cent. of phosphorus, and the remainder copper. Second, its manufacture by melting a bath of copper, adding to it aluminium in the proportion stated, the bath being covered with a layer of palm oil to prevent oxidation, and then adding a small proportion of phosphorus.

Cowles Bros. in their pamphlet give the following tests of the strength of aluminium-silver castings:—

	Tensile strength, pounds.	Elongation, per cent.
5 p. ct. Al bronze, 1 part; Ni 2 parts	79,163	33.0
" " 4 parts; Ni 1 part	118,000	...
German silver without aluminium	44,242	24.0
" " with "	92,849	1.0

Solders for Al Bronze.—Cowles give the following jeweller's solder for aluminium bronze:—

---

* U. S. Pat., 303,236. Aug. 1884.

Hard solder for 10 per cent. bronze—

Au	88.88
Ag	4.68
Cu	6.44

Middling hard solder for 10 per cent. bronze—

Au	54.40
Ag	27.00
Cu	18.00

Soft solder for Al bronze—

Cu 70 per cent. / Sn 30 "	Bronze	14.30
	Au	14.30
	Ag	57.10
	Cu	14.30

Silicon and Aluminium Bronze.—Cowles Bros. have, by reducing fire clay in presence of copper, obtained alloys of aluminium, silicon, and copper. This alloy is white and brittle if it contains over 10 per cent. of aluminium and silicon together. With from 2 to 6 per cent. of these in equal proportions, the alloy is stronger than gun metal, is very tough, does not oxidize when heated in the air, and has a fine color. Cowles report that a silicon-aluminium bronze wire has shown a tensile strength of 200,000 pounds, a strength hitherto unprecedented in any metal.

Aluminium and Iron.

Tissier Bros., 1858: " An alloy of aluminium and iron with 5 per cent. iron was made by placing very fine iron wire with fragments of aluminium in a crucible containing melted NaCl. Under these circumstances the iron could not oxidize, and the alloy was easily formed. We have in this way been able to discover that small quantities of iron give to aluminium the property of crystallizing, and much impair its malleability. When aluminium has become low in price, it will be interesting to see what qualities it can communicate to iron as cast iron or steel, introduced in large or small quantities. Iron raises the fusing point of the aluminium, for we have melted aluminium free from iron on a plate of aluminium containing 4 to 5 per cent. iron.

Deville, 1859: . " Iron and aluminium combine in all proportions. These alloys are hard, brittle, and crystallize in long needles, when the proportion of iron reaches 7 or 8 per cent. The alloy containing 10 per cent. iron much resembles sulphide of antimony. It liquates, however, with some facility, leaving a less fusible skeleton, while less ferruginous aluminium runs down. But this method of purifying aluminium is not exact. The presence of a large quantity of iron in aluminium alters both its chemical and physical properties."

Rogers:[*] " By melting a steel high in carbon with aluminium, alloys of steel and aluminium may be obtained. I have one containing 6.4 per

---

[*] Moniteur Industriel, 1859, No. 2379.

cent. of the latter. I melted 67 parts of this alloy with 500 of steel, so that the resulting steel contained 0.8 per cent. aluminium. This metal had the qualities of the best Bombay wootz. A small per cent. of aluminium makes steel hard, strong, and brittle, a larger quantity makes it very dense, without impairing its peculiar polish or detracting from its qualities."

Fremy, 1883: "Aluminium unites with iron with the greatest facility. To form an alloy it is sufficient to stir a rod of iron in melted aluminium, when it covers itself with a layer of aluminium and takes on the aspect of being amalgamated. The alloy with 5 per cent. iron is hard, brittle, and more difficult to fuse than aluminium. The 7 per cent. iron alloy possesses the same properties, with a crystalline structure. The 10 per cent. iron alloy, according to Deville, resembles sulphide of antimony, Sb^2S^3. On the other hand, M. Debray affirms that 7 to 9 per cent. of iron in aluminium causes no appreciable change in its properties. By melting ten parts aluminium, five parts ferric chloride, and twenty parts KCl and NaCl, Michel obtained on cooling a mass which, treated with very dilute sulphuric acid, left six-sided prisms having the color of iron and the formula Al^2Fe, containing 51 per cent. Fe.[*] Calvert and Johnson obtained the alloy Al^2Fe^3,

---

[*] Ann. Chem. und Pharm. 115, 102.

ALLOYS OF ALUMINIUM. 283.

containing twenty-four per cent. aluminium and seventy-six per cent. iron, which was unalterable in moist air (see p. 210). The alloy $AlFe^4$, containing 10.8 per cent. aluminium, has been prepared by melting two parts aluminium, five parts sheet iron, and one part of chalk. It is easily worked and rolled, but rusts on contact with the air."

Mierzinski: " A few per cent. of aluminium is useful in making cast steel, to which it imparts greater hardness and a bright silver-like polish; 0.8 per cent. aluminium gives steel all the qualities of best Bombay wootz, and objects made of it, treated with dilute sulphuric acid, give the undulating markings generally found only on Damascus steel. Stoddart and Faraday found in wootz steel 0.013 to 0.690 per cent. of aluminium. An alloy of 24.5 per cent. aluminium is silver white, extraordinarily hard, and does not rust in the air.

Mitis castings:* " The subject of the use of aluminium in wrought-iron castings was discussed at the meeting of the American Society of Mining Engineers, Pittsburgh meeting, on Feb. 16. It was described by the inventor, Mr. Peter Ostberg, of Stockholm, and as his paper has not yet appeared we give a few particulars which may be of interest.

" Wrought-iron scrap is melted in plumbago crucibles in a special reverberatory furnace fired

* Eng. and Mining Journal, Feb. 27, 1886.

with petroleum. The crucible is covered, while a hole in a cover corresponds with and is directly under a hole in the roof of the furnace. Wrought iron fuses at about 4000° F., and it would be necessary to heat it far beyond its point of fusion before it would be fluid enough to cast into fine moulds and to make it possible to handle it before it would solidify. Now it is in this superheating that the iron absorbs gases, and consequently it is impossible to make solid castings in this way. In order to obviate this difficulty, Mr. Ostberg has made use of the well-known fact that certain alloys of metals possess a fusing point much less than that of the metals composing them, among which aluminium alloys are very noticeable. In making wrought-iron or mitis castings a very small quantity of aluminium, about 0.05 per cent., is added to the charge in the crucible the moment it has been melted. The charge is about sixty pounds. The aluminium is added in the form of an iron-aluminium alloy containing 7 to 8 per cent. of aluminium. The fusing point of the whole is at once lowered some 500 degrees, and the charge being then nearly 500 degrees above its new fusing point becomes extremely fluid and can be cast into the finest moulds; while the great difference between its temperature and its reduced fusing point gives all the time necessary for manipulating it without danger of solidifying. This extreme fluidity allows the ready escape of gases which would otherwise make the casting porous, and the result appears to be a remarkably fine, solid, and tough casting of wrought iron.

"These mitis castings are said to be 30 to 50 per cent. stronger than the iron from which they are made; but, although aluminium undoubtedly increases the strength of most metals with which it alloys, it is not credited with the increase of strength in this case; for it is said that after hammering, the mitis metal loses its increase in strength and returns to the fibrous appearance and strength of the original iron.

"The alloys of iron and steel with aluminium have long been known, and reference is made to the addition of such an alloy to steel by Faraday, only a few years after the discovery of aluminium; but this application to wrought iron castings appears to be new and is certainly very interesting.

"The alloy used by Mr. Ostberg at his works in Worcester, Mass., is made by the Cowles Electric Smelting Co., and contains 6 to 8 per cent. aluminium and 1 to 1.25 per cent. of silicon. It costs about forty cents per pound, but as only 0.05 per cent. aluminium is required in the iron, the addition to its cost is very slight. This utilization of the well-known property of aluminium to lower the fusing point of the iron is a very neat and clever application of a curious phenomenon, and it is said to succeed very well. Whether it will also facilitate the making of small steel castings is not stated, but it would probably in this case make the metal more fluid and obviate the necessity of using those extremely high heats which are necessary to cause the steel to melt and run well into the moulds."

Mr. Ostberg sent a note to the 'Engineering and Mining Journal,' stating that he used only a small sample of Cowles' alloys, but that he uses almost altogether a 7 to 8 per cent. aluminium alloy, made in Sweden by a very simple and cheap patented process, which consists in adding clays in iron smelting.

The following note, bearing on this subject, is from Watts: "The 'London Mining Journal' states that if common kaolin is added to iron when being smelted in a crucible to convert it into steel, an improved product is the result." Aside from this note, the author has been unable to find any reference to the process suggested by Mr. Ostberg.

Mr. Sellers, of Philadelphia, remarked after the reading of Dr. Hunt's paper on the Cowles furnace at the Washington meeting of the National Academy of Sciences, April, 1886 (see p. 196), that he had made a series of experiments on the use of aluminium with iron in casting, and obtained what is technically called a "dead melt" in two or three minutes, instead of an hour as required by previous methods. The result is very fine castings, and without the flaws which so often vex the founder.

A company has been incorporated in New Jersey within the last month to regulate the use and sell rights to use Mr. Ostberg's patents. Mr. Fritz, of the Bethlehem Iron Works, is one of the heads of the company, which includes other prominent Bethlehem capitalists.

ALUMINIUM AND ZINC.

Tissier Bros., 1858: "An alloy with 10 per cent. aluminium is brittle, has the appearance of zinc, is more fusible than aluminium, and less so than zinc. An alloy with 25 per cent. aluminium has a fine even grain, and is still more fusible than aluminium and less so than zinc. An alloy with 50 per cent. did not appear to be homogeneous; heated on an aluminium plate it separated into a fusible portion and a part which did not melt till the plate did. These alloys have been tried as solders for aluminium, and so far have succeeded better than any other alloys, but, unfortunately, when melted they are thick and cast with difficulty, so much so that it is necessary to spread them over the joint as a plumber does when he wipes the joints of lead pipes. Joints thus made stand hammer blows or rough usage very poorly."

Deville, 1859: "The alloys of aluminium and

zinc are brittle, at least unless the zinc is in small proportion. Several specimens of zinciferous aluminium were put into commerce by a singular accident; the retorts used for making the aluminium were made at the Vielle Montagne Zinc Works, and having in their mixture some ground-up old zinc retorts, the new retorts contained zinc, which got into the aluminium and altered its properties in a very evident manner. Some analyses of this metal having been made in England, some asserted that French aluminium was only an alloy to which zinc gave a fusibility which might be wanting in pure aluminium. The alloys of zinc and aluminium have been used in experiments to solder aluminium solidly, but so far with little success. Zinc unites easily with the aluminium, altering its properties when exceeding a few per cent."

Kerl and Stohman, 1874: "Zinc and aluminium melted together in atomic proportions under a cover of NaCl and KCl unite with incandescence, forming a silver-white, very brittle, crystalline alloy, with a specific gravity of 4.532."

Fremy, 1883: "The alloys of zinc and aluminium are employed to solder aluminium. They will take a very fine polish. The alloy with three per cent. zinc is yet malleable, but that with thirty per cent. aluminium is white, crystalline, and very brittle."

ALUMINIUM AND TIN.

Tissier Bros., 1858: An alloy with 3 per cent. tin is very brittle, a little more fusible than aluminium. It was made by combining the metals under a cover of NaCl, then remelted alone, and cast. Its grain is very fine and crossed, and it breaks at the first blow of the hammer. If tin, even in small quantities, injures the qualities of aluminium, the latter, on the contrary, gives to tin hardness and tenacity, if it is not present in too large an amount. The alloy with 5 per cent. aluminium possesses these desirable qualities. The alloy with 10 per cent. is not homogeneous, for it arranges itself in the ingot in two layers, an upper brittle one, a little more fusible than aluminium, another lower one containing nearly all the tin but rendered harder and less fusible by a small quantity of aluminium. These alloys have been used to solder aluminium because of their fusibility and the ease with which they adhere to a clean surface, but they have the same inconveniences as the zinc solders, they run thick and are fragile.

Deville: Tin unites easily with aluminium, altering its properties as soon as the proportion has passed a few per cent. These alloys may be used to solder aluminium, but they answer imperfectly.

Kerl and Stohman: The aluminium-tin alloys with over 30 per cent. of the former are silver-white, but porous and brittle. The 19 per cent. and especially the 7 per cent. aluminium alloys are, on the contrary, malleable and workable at a red heat.

Fremy: A small quantity of tin renders aluminium brittle, but a small per cent. of aluminium alloyed with tin renders it harder and more elastic. Such an alloy, besides being easy to work, may advantageously replace tin in many of its uses. The alloys most recommended are those with 5, 7, and 19 per cent. of aluminium.

Mierzinski: Aluminium and tin unite in certain proportions, but the tin will not combine with more than 7 per cent. of aluminium, for the 10 per cent. aluminium alloy is no longer homogeneous but on cooling liquates away a more fusible and leaves a less fusible alloy, the latter being richer in tin. An alloy with 3 per cent. aluminium is harder than tin and less acted on by acids. The 7 per cent. aluminium alloy is especially recommended as being easy to work, capable of being polished, but possessing the drawback that it cannot be melted without a part of the tin separating from the aluminium.

M. Bourbouze,[*] a French physicist, employs an alloy of aluminium and tin for the interior parts of optical instruments, in place of brass. The alloy contains 9 per cent. of tin. It is white, like

---

[*] Iron Age, July 29, 1886.

aluminium, and has a density of 2.85. This lightness is of great advantage. It can be soldered as easily as brass, without special apparatus, and is more resistant to reagents than aluminium. It would be very useful for electrical instruments, especially those of a portable character.

ALUMINIUM AND LEAD.

Tissier: As Deville has remarked, these two metals have such little tendency to combine that there may be recovered intact in the bottom of an ingot of aluminium any small pieces of lead which may accidentally have got into the metal.

Deville: Lead unites only imperfectly with aluminium. However, an alloy may exist in certain proportions, especially at the temperature necessary to the cupellation of aluminium. The cupellation of aluminium with lead is quite possible.

Kerl and Stohman: Aluminium does not unite with lead.

Mierzinski: Aluminium and lead do not unite. By melting the metals together and cooling down, they are found separated from each other, the aluminium above and the lead beneath. This property suggests the possibility of using it to separate silver from work lead, as soon as its price allows.

Aluminium and Antimony.

Tissier: Aluminium also appears to have as little tendency to unite with antimony as with lead; we did not succeed in getting a homogeneous alloy of the two metals.

Kerl and Stohman: Aluminium does not unite with antimony.

Aluminium and Bismuth.

Tissier: The combination of these two metals takes place easily and gives rise to very fusible alloys, which oxidize very rapidly when melted. They are also very alterable in the air at ordinary temperatures when the bismuth is in large per cent. However, these metals do not appear able to unite in all proportions, as the following experiment seems to prove. We melted together 10 grms. aluminium and 20 grms. bismuth. The combination took place under NaCl, and although it was stirred carefully the button appeared to be of two layers, the lower one composed of almost pure bismuth, and the upper one more malleable, detachable from the lower by a blow of the hammer, and weighing 13.45 grms. Supposing that this latter contained all the aluminium (there appeared to be none in the other layer), then the alloy was composed of nearly 75 per cent. aluminium and 25 per cent. bismuth. Thus the aluminium did not appear able to take up over 25 per cent. of bismuth. The alloy containing 10 per cent. bismuth is hard, malleable, takes a fine polish, is unattacked by nitric acid, and not blackened by sulphuretted hydrogren. We say it was malleable because it could be worked to a certain extent under the hammer, and we thought it could be easily drawn out, but in spite of frequent annealings it split in all directions and we had to stop working it. We tried, by diminishing the per cent. of bismuth, to take away this bad quality, and to this end prepared alloys with 5, 3, 2.5, and 0.5 per cent. of the latter, but without obtaining satisfactory results.

Watts: One-tenth of one per cent. of bismuth renders aluminium so brittle that it cracks under the hammer after being repeatedly annealed.

ALUMINIUM AND NICKEL.

Tissier: The alloy with 50 per cent. nickel was made by melting together the metals in equal proportions under $NaCl$; the heat evolved was sufficient to raise the mass to incandescence. This alloy remained pasty at the temperature of melting copper. It is so brittle that it pulverizes under the hammer. By melting proper proportions of this alloy with more aluminium, an alloy with 25 per cent. nickel was produced. This is less fusible than aluminium, and as brittle as the 50 per cent. alloy. By melting some 25 per cent. nickel alloy with aluminium, a 5 per cent. nickel alloy was obtained. This is much less brittle than the preceding, but is still very far from being easy to work. From the 5 per cent. alloy one with 3 per cent. was made. With this amount of nickel the aluminium acquired much hardness and rigidity, and was easy to work. A curious fact with this alloy is that it may be melted on a plate of aluminium, showing its fusion point to be less than that of pure aluminium, the reverse effect to what iron produces, which if present in the same proportion would diminish the fusibility of the aluminium. To sum up, the action of nickel on aluminium is much analogous to that of iron, for nickel, like iron, produces crystalline alloys with aluminium, and if employed with care gives to it certain desirable qualities such as hardness, elasticity, etc.

Mierzinski: To alloy aluminium with nickel a certain limiting quantity of nickel must not be exceeded. When the latter is present less than 3 per cent., it behaves similarly to iron in improving the qualities of the aluminium in many ways, especially in hardness and elasticity. More than 3 per cent. makes the aluminium brittle and unworkable.

Argentan has a beautiful color, and takes a high polish, it contains—

Cu	70
Ni	23
Al	7

ALLOYS OF ALUMINIUM.

Minargent contains—

Cu	100
Ni	70
Sb	5
Al	2

ALUMINIUM AND SILVER.

Tissier: Silver is the metal which seems most useful in improving aluminium. Five per cent. silver gives to aluminium elasticity which is wanting in pure aluminium, increases its hardness and its capability of being polished, and does not injure its malleability. We have sold a quantity of these alloys, the properties of which we will describe. All the alloys up to 50 per. cent of silver are more fusible than aluminium, the fusibility increasing with the amount of silver. The alloy with 33 per cent. silver is fusible enough to serve for a solder; but, like the alloys of aluminium with zinc and tin, it casts with difficulty and makes a brittle joint. With 10 per cent. silver the aluminium will not stand under the hammer. The 50 per cent. alloy breaks like those of copper. The presence of silver in aluminium can always be recognized by the action of the alloy on a moderately concentrated solution of caustic potash. Aluminium whitens in this solution, but, if it contains silver, this being exposed by the dissolving away of the aluminium gives the surface a black color. By introducing 5 per cent. of aluminium into silver the latter acquires the hardness of silver coin, the alloy takes a beautiful polish, does not contain as alterable a metal as copper, and contains 95 per cent. of silver instead of 90. This alloy is easily distinguished from the alloy into which copper enters by the test with nitric acid, which whitens instead of blackening it.

Deville: A few per cent. of silver will take away from aluminium all its malleability. However, the alloy with 3 per cent. is used by M. Christofle for casting objects of art; and the alloy with 5 per cent. to make knife blades, and it may be worked like pure aluminium. It has, moreover, the color and lustre of silver, and is not tarnished by sulphuretted hydrogen.

Kerl and Stohman: According to Hirzel, the alloy containing 20 per cent. aluminium is very porous, silver-white, tarnishing in the air, sp. gr. 6.733. $AlAg^2$, containing 11.11 per cent. aluminium is also silver-white, a little porous, tarnishes in the air, sp. gr. 8.744. $AlAg^4$, containing 5.9 per cent. aluminium is pure silver-white, very malleable, forgeable, tarnishing in the air, sp. gr. 9.376.

Fremy: The alloys of aluminium and silver are easy to form by direct fusion of the two metals; their hardness is generally superior to that of aluminium, but, nevertheless, they are quite as easy to work, and in some cases more fusible than it. Debray states that the 50 per cent. alloy is as hard as bronze.

Mierzinski: Five per cent. of silver makes aluminium elastic and as hard as coin silver, but not brittle. This alloy is workable like pure aluminium, takes a fine polish, is light, not magnetic, does not rust, and has the color of pure silver, whose place it can take for many purposes. However, the assertion that this or any other alloy of silver and aluminium is not attacked by hydrogen sulphide is incorrect and untenable, since, according to careful experiments, these alloys are attacked quicker and more actively by it than pure silver. This alloy is used for watch-springs, dessert-spoons, etc., on account of its hardness and elasticity. The alloy with 3 per cent. silver has a very fine silver color. The 50 per cent. alloy is as hard as bronze, but so brittle that it cannot be pressed; all the alloys with over 10 per cent. silver up to the 50 per cent. alloy are brittle and cannot be worked with a hammer.

"Tiers Argent" is an alloy of two-thirds aluminium and one-third silver, which was made homogeneous at first with some difficulty but is now easily made. Spoons, forks, and salvers of this alloy leave nothing to be desired. It possesses a hardness superior to silver, and can be easily engraved.*

Cowles Bros. state that what is generally known and sold as aluminium silver is an alloy of alu-

* Chem. News, xvi. 289.

minium, nickel, and copper; or, in effect, it is aluminium added to German silver. The great advantage of this metal is that it will keep its beautiful white lustre for all time, and permits of objects being made from it in an enduring and substantial manner. It requires no plating of any kind.

Aluminium and Gold.

Tissier: Aluminium endures a large quantity of gold without its ductility being impaired. We have prepared an alloy with 10 per cent. of gold which works at a red heat as well as aluminium, is a little harder but scarcely polishes any better than it. Its color, for some cause, is darkish brown, like that of tin lightly sulphurized. The alloy containing 15 per cent. gold can no longer be forged. As to the effect of small quantities of aluminium on gold, 5 per cent. of it gives to the latter a white color and makes it brittle as glass.

Fremy: The alloy with one per cent. aluminium possesses the color of "green gold;" it is very hard but yet malleable. The alloy with 10 per cent. aluminium is white, crystalline, and brittle; the alloy with 5 per cent. is brittle as glass.

Mierzinski: Aluminium can take up as much as 10 per cent. of gold without its malleability decreasing. This alloy can be forged, but not well polished. The color of the gold has entirely disappeared, seeming to have no effect on the aluminium.

ALUMINIUM AND PLATINUM.

Tissier: Aluminium unites with platinum with great ease, forming with it alloys more or less fusible according to the proportions of aluminium. Five per cent. of platinum makes an alloy not malleable enough to be worked; it is probable that by diminishing the amount of platinum a suitable alloy might be produced. In color it approaches that of gold containing 5 per cent of silver.

ALUMINIUM AND CADMIUM.

Deville: Cadmium unites easily with aluminium. The alloys are all malleable and fusible, and may be used to solder aluminium, though imperfectly.

ALUMINIUM AND BORON.

Deville: "By melting aluminium with borax, boracic acid, or fluo-borate of potassa, an alloy very rich in boron was obtained. This alloy, like siliceous aluminium, possesses the singular property that the boron diminishes all its useful qualities. The alloy is very white, only able to bear slight bending, and tears under the rolls. It exhales a very strong odor of hydrogen silicide, SiH^4, without doubt due to the silica of the vessel which was attacked at the same time as the borax. M.

Wöhler and I have shown that the boron may be extracted from this alloy in two different forms, the graphitoidal and the diamantine boron." Deville gives at the end of his volume on aluminium the mode of preparation of this diamantine boron.

Aluminium and Carbon.

Deville: I was not able, by any effort I made, to combine carbon with aluminium. On decomposing carbon tetrachloride, CCl^4, by aluminium, there is formed ordinary carbon, while the aluminium which remains has undergone no change.

Cowles: Specimens of alloys of aluminium and carbon, yellow and crystalline, have been exhibited, which were made in the Cowles furnace. (See p. 195.)

Aluminium and Gallium.

Watts: Lecoq de Boisbaudran makes the following remarks: "If the proportion of aluminium is to be considerable, the two metals are melted together at dull redness. The alloys thus obtained remain brilliant, and do not sensibly absorb the oxygen of the air in their preparation. After cooling they are solid but brittle, even when the excess of aluminium has raised the melting point to incipient redness. They decompose water in the cold, but better at 40°, with rise of temperature, evolution of hydrogen, and formation of a chocolate-brown powder, which is ultimately resolved into white flakes of alumina."

ALUMINIUM AND TITANIUM.

"Wöhler* fused in a clay crucible 10 grms. of titanic acid, 30 grms. cryolite, 15 grms. each of NaCl and KCl, and 5 grms. of aluminium. This was kept at the melting point of silver for one hour and then opened. The aluminium had become lammelar, and when dissolved in caustic soda left a quantity of brilliant, crystalline plates, found to be a compound of aluminium, titanium, and silicon. The elements of the compound appear to be able to unite in various proportions. Its density was 3.3. It was infusible before the blowpipe, but heated to redness in chlorine it burnt, giving chlorides of the three metals present. Another experiment, heated only to the melting point of nickel, gave a white compound richer in silicon, sp. gr. 2.7.

Alloys† of aluminium with wolfram, molybdenum, and manganese were made by Michel in Wöhler's laboratory, on which the following report is made:—

* Chem. News, 1860, p. 310.
† Ibid.

Aluminium and Tungsten.

Al^4W was made by fusing together 15 grms. tungstic acid, 30 grms. cryolite, 15 grms. each of KCl and NaCl, and 15 grms. aluminium, at a strong red heat. The excess of aluminium was removed from the regulus by HCl. The alloy is an iron-gray powder, crystalline, single crystals were several millimetres long, brittle and hard rhombic prisms. Sp. gr. 5.58. Hot caustic soda extracts all the aluminium from these crystals, leaving behind pure tungsten.

Aluminium and Molybdenum.

Molybdic acetate is dissolved in hydrofluoric acid, the solution evaporated to dryness, and the residue mixed with cryolite, flux, and aluminium, in the same proportions as given for tungsten. Excess of aluminium is dissolved from the product with caustic soda, and there remains a black, crystalline powder consisting of iron-gray rhombic prisms, soluble in hot nitric or hydrochloric acid, and consisting entirely of aluminium and molybdenum.

Aluminium and Manganese.

We fused together 10 grms. anhydrous manganese chloride, 15 grms. each of KCl and NaCl, and 15 grms. aluminium. The excess of aluminium was removed by HCl. There remained a dark-gray, crystalline powder consisting of square prisms, specific gravity 3.4. Dilute caustic soda extracts all the aluminium from these, leaving the manganese.

ALUMINIUM AND SODIUM.

Deville: Aluminium unites easily with sodium, especially in small proportions. From this it follows that the properties of the metal made carelessly by using sodium are completely altered. The last traces of sodium can be removed only with great trouble, especially when the aluminium has been produced in presence of fluorides, because of the marked affinity of aluminium for fluorine at the temperature at which aluminium fluoride, Al^2F^6, commences to volatilize.

Fremy: Aluminium easily combines with sodium. If the combination contains 2 per cent. of sodium, it easily decomposes water, which circumstance gave cause to the notable loss of aluminium when it was first being manufactured.

ALUMINIUM AND NITROGEN.

Dr. Hunt,[*] in reading a paper on the Cowles furnace (see p. 196), showed a specimen of a peculiar alloy believed to consist entirely of aluminium and nitrogen.

[*] Washington Meeting, Nat. Acad. Sciences, April, 1886.

APPENDIX.

NATIVE SULPHATE OF ALUMINA.

IN the summer of 1884, a large deposit of rock called "native alum" was discovered on the Gila River, Sorocco Co., New Mexico, about two miles below the fork of the Little Gila and four miles below the Gila Hot Springs. The deposit is said to extend over an area one mile square and to be very thick in places. The greater part of the mineral is impure, as is usual with native occurrences, but it is thought that large quantities are available. A company formed in Sorocco has taken up the alum-bearing ground. Through the kindness of Mr. W. B. Spear, of Philadelphia, I was enabled to get a specimen of it.

It is white, with a yellowish tinge. On examining closely it is seen to consist of layers of white, pure-looking material arranged with a fibrous appearance at right angles to the lamination. These layers are about one-quarter of an inch thick. Separating them are thin layers of a material which is deeper yellow, harder and more compact. The whole lump breaks easily and has a strong alum taste. On investigation, the fibrous material was found to be hydrated sulphate of alumina, the harder material sulphate of lime.

It is probable that this deposit was the bed of a shallow

lake in which the alum-bearing water from the hot springs concentrated and deposited the sulphate of alumina. Periodically, or during freshets, the Little Gila, flowing through a limestone country, bore into this lake water containing lime, which, meeting the $Al^2(SO^4)^3$ solution, immediately caused a deposit of $CaSO^4$. When the dry season came, the Little Gila dried up, the deposit of alum was made, and thus were formed the succession of layers through the deposit.

Analysis showed 7 to 8 per cent. insoluble material, and the remainder $Al^2(SO^4)^3.18H^2O$. A small amount of iron was present.

Decomposition of Cryolite.

According to a patent given to F. Lauterborn (see p. 206), cryolite can be decomposed by boiling with water; sodium fluoride going into solution and aluminium fluoride remaining as residue.

To test the accuracy of this statement, I boiled 250 grms. of the mineral in 5 litres of water for 3 hours. The solution was filtered hot and evaporated to dryness. There was no residue. The material on the filter appeared to be undecomposed cryolite.

The experiment does not prove that the decomposition is impossible, but makes it appear extremely improbable.

American Aluminium.

I bought some of Mr. Frishmuth's aluminium from Bullock & Crenshaw, Philadelphia. The surface was slightly whitened by oxidation, resembling, though not to such a degree, the oxidation which takes place on slabs of zinc when exposed to the air. The wire was not perfectly smooth, being at places slightly rough and scaly. It was quite soft and malleable. Its color was nearly white, but with a slight blue tinge, which, if intensified, would have made it resemble zinc more than any other metal.

Duplicate analyses of it gave me the following results:—

Si	0.65	0.56
Fe	1.94	1.87
Al (by diff.) . . .	97.41	97.57

After making these analyses, I came across the analysis of Mr. Frishmuth's metal given on p. 53.

Specific Gravity of Aluminium.

The sp. gr. of the metal whose analysis was just given I determined very accurately on a Becker balance. Compared with water at 4° C., it was 2.735. I wished to see if this would correspond to the sp. gr. calculated from the analysis. The data were as follows:—

	Average sp. gr.	Average per cent. present.	Products.
Si . . .	2.34	0.60	0.014
Fe . . .	7.7	1.80	0.138
Al . . .	2.67	97.60	2.605

Calculated sp. gr. 2.757

The correspondence being so close has suggested that the sp. gr. of commercial aluminium, carefully taken, gives the approximate amount of iron present; for the sp. gr. of silicon is so near to that of aluminium, that 10 per cent. of the former, an amount never found in commercial aluminium at present, would only affect the sp. gr. 0.03.

So then, within the limits usually found in commercial aluminium, *i. e.*, silicon less than 5 per cent. and iron anywhere less than 10 per cent., a careful determination of the sp. gr. should, by a little calculation, give the amount of iron present within a limit of error of 0.5 per cent. at most, thus saving a wet determination of iron.

AMALGAMATION OF ALUMINIUM.

Wishing to observe the effect of mercury on metallic aluminium, I took a clean, bright piece of aluminium foil of Mr. Frishmuth's make, and put on it a small globule of mercury, which I rubbed in with the finger. Almost immediately a white powder appeared and the foil felt warm from the heat generated. On brushing away this powder, the foil underneath appeared white and unattacked. By letting the mercury remain on the foil, it very soon eat a hole through it. Compare with p. 261.

It thus appears that mercury unites with a clean surface of aluminium, forming an amalgam, and the aluminium in the amalgam oxidizes in the air to alumina. The question arises, why does aluminium oxidize so easily? We know how the properties of this metal depend much on its state of division; its foil will burn in the air, whereas the metal in bulk will not. The mercury serves, as it amalgamates the aluminium, to draw apart even the molecules of the metal, and so this extremely minute, even molecular division of the aluminium permits it to exhibit in an intensified degree the principle just stated, which was illustrated by the burning of the foil, *i. e.*, the finer its state of division the more easily is it acted on by oxidizing agents. Translating this into the language of chemical affinities, in metallic aluminium the atoms are united two by two by a mutual exchange of affinities, and the oxygen of the air is not able to break this molecular bond at ordinary temperatures. But by the intervention of the mercury this bond is broken, and the atoms of aluminium become united with atoms of mercury, which weakens the bond holding the molecule together. The strong affinity which oxygen has for aluminium is now able to break up the new molecule, the metal is rapidly oxidized and the mercury set free.

Reduction of Alumina.

I experimented on reducing alumina by carbon in presence of copper. (See p. 213.) I took for a charge—

40 grms. . . . . CuO and Cu.
5 " . . . . Al^2O^3.
5 " . . . . Charcoal.

These were intimately mixed and finely powdered, put in a white-clay crucible and covered with cryolite. The whole was slowly heated to bright redness, and kept there for two hours. A bright button was found at the bottom of the crucible. This button was of the same sp. gr. as pure copper, and a qualitative test showed no trace of aluminium in it.

This is the same result that other experimenters have reached, and the conclusion seems to be that the process gives no practical results.

Production and Reduction of Aluminium Sulphide, Al^2S^3.

Until the researches of M. Fremy, no other method of producing Al^2S^3 was known save by acting on the metal with sulphur at a very high heat. Fremy was the first to open up this new field, and it may be that his discoveries will yet be the basis of successful industrial processes. Fremy is often quoted in connection with Al^2S^3, and in order to understand just how much he discovered we here give all that his original paper contains concerning this sulphide.*

"We know that sulphur has no action on silica, boric oxide, magnesia, or alumina. I thought that it might be possible to replace the oxygen by sulphur if I introduced or intervened a second affinity, as that of carbon for oxygen. These decompositions produced by two affinities are very frequent in chemistry, it is thus that carbon and chlorine, by acting simultaneously on silica or alumina, produce silicon or aluminium chloride, while either alone could not decompose it; a similar case is the decomposition of chromic oxide by carbon bisulphide, producing chromium sesquisulphide. Reflecting on these relations, I thought that carbon bisulphide ought to act at a high heat on silica, magnesia, and alumina, producing easily their sulphides. Experiment has confirmed this view. I have been able to obtain in this way almost all the sulphides which until then had been produced only by the action of sulphur on the metals.

"To facilitate the reaction and to protect the sulphide from the decomposing action of the alkalies contained in the porcelain tube which was used, I found it sometimes useful to mix the oxides with carbon and to form the mixture into bullets resembling those employed in the preparation of Al^2Cl^6. I ordinarily placed the bullets in little

---

* Ann. de Chem. et de Phys. [3] xxxviii. 312.

ALUMINIUM SULPHIDE. 311

carbon boats, and heated the tube to whiteness in the current of vaporized carbon bisulphide. The presence of divided carbon does not appear useful in the preparation of ▓▓▓▓▓ phide.

The Al^2S^3 formed is not volatile; it remains in the carbon boats and presents the appearance of a melted vitreous mass. On contact with water it is immediately decomposed.

$$Al^2S^3 + 3H^2O = Al^2O^3 + 3H^2S.$$

"The alumina is precipitated, no part of it going into solution. This precipitated Al^2O^3 is immediately soluble in weak acids. The clear solution, evaporated to dryness, gives no trace of alumina. It is on this phenomenon that I base a method of analysis as will be seen below.

"Analysis of the product. Al^2S^3 being non-volatile, it is always mixed with some undecomposed alumina. It is, in fact, impossible to entirely transform all the alumina into Al^2S^3. I have heated less than a gramme of alumina to redness five or six hours in carbon bisulphide vapor, and the product was always a mixture of Al^2O^3 and Al^2S^3. The reason is that the sulphide being non-volatile and fusible coats over the alumina and prevents its further decomposition. The Al^2O^3 thus mixed with the Al^2S^3, and which has been exposed to a red heat for a long time, is very hard, scratches glass, and is in grains which are entirely insoluble in acids. By reason of this property I have been able to analyze the product exactly, for on treating the product with water and determining on the one hand the sulphuretted hydrogen evolved, and on the other the quantity of soluble alumina resulting, I have determined the two elements of the compound. One gramme of my product contained 0.365 grm. of Al^2S^3, or 36.5 per cent., the remainder being undecomposed alumina." The composition of this Al^2S^3 was—

Al	. . .	0.137 grm.	= 37.5	per "
S	. . .	0.228 "	= 62.5	"
		0.365 "	100.0	"

The formula Al^2S^3 requires—

Al . . . . . . 36.3 per cent.
S . . . . . . 63.7 "

The above is the substance of Fremy's remarks on Al^2S^3. The next investigation in this field was made by Reichel. His paper is on the sulphides of magnesium and aluminium and he proceeded in methods so similar with both metals that he sometimes describes a process only for magnesium sulphide, MgS, with details, and merely states his results in working the same way for Al^2S^3, which will account for the frequent allusion in his paper to MgS. The paper is very lengthy, but only what bears directly on the subject in hand is extracted.

" I wished* to obtain more definite knowledge of MgS and Al^2S^3, and I also had a practical end in view; for, depending on the small affinity of sulphur for magnesium and aluminium, I hoped, if not to isolate them from the sulphide, at least to try the possibilities of this method." (As preliminary, Reichel here gives extended remarks on the behavior of aluminium and magnesium towards sulphur, and a description of the sulphides.)

" Al^2S^3 appears yellow, at least that made from the metal and sulphur with exclusion of air always has this

* Jrnl. fr. Prak. Chem. xi . 55.

ALUMINIUM SULPHIDE. 313

color. It is only by heating the metal with sulphur vapor with admittance of air that the product is of a darker colo▓▓▓▓▓ experiments to determine if the preparation of ▓▓▓▓ and MgS was not possible in the same way as K^2S, Na^2S, and BaS are made. For example, by melting potassium oxide with sulphur some K^2S is formed. However, on doing this with alumina and magnesia, it became evident that magnesium and aluminium have less affinity for sulphur than for oxygen, and the experiment failed. But, matters were changed when a reducing agent was introduced with the sulphur. If a mixture of carbon, magnesia, and sulphur are heated, MgS results, which by treating the product with water goes into solution unchanged. Alumina under similar treatment gave no Al^2S^3. In place of carbon as a reducing agent I next used hydrogen. By igniting magnesia in a stream of hydrogen and sulphur vapor some little MgS was formed, but the mass of the magnesia was unchanged. The next step was to substitute sulphuretted hydrogen for hydrogen and sulphur separately, but only a little MgS was formed in this way.

"Since the sulphates of calcium and barium are reduced to sulphides with very little trouble, it appeared probable that magnesium sulphate, $MgSO^4$, should be convertible into MgS by a reducing agent. The attempts to do this were unsuccessful. I heated $MgSO^4$ in vapor of ammonium sulphide, but it underwent no change. Since according to Stammer* K^2SO^4, $CaSO^4$, and $BaSO^4$ may be reduced to sulphides by carbonic oxide, CO, I tried to

* Pogg. lxxxii. 135.

reduce $MgSO^4$ by this means. The following reaction apparently took place—

$$MgSO^4 + 4CO = MgO + COS, + 3CO^2.$$

"I then took pure magnesia, filled a porcelain with it, and passed carbon bisulphide vapor through it. The apparatus was first filled with hydrogen, then as soon as the tube was bright red the carbon bisulphide flask was warmed, and sulphuretted hydrogen and carbonic oxide began to issue from the tube. The heating was continued till carbon bisulphide condensed in the outlet tube, then the fire was removed and hydrogen passed through the tube till it was cold. The MgS resulting was of a gray color, not melted, but as a crumbly powder. The reaction which took place was probably

$$MgO + 2CS^2 + 6H = MgS + 3H^2S + CO + C.$$

"The carbon was left with the MgS; and, to get rid of it I heated the tube up as before but passed hydrogen and carbonic oxide through, when the hydrogen took up the carbon forming probably some hydrocarbon.

"Than* says that carbon oxysulphide, COS, is formed by leading carbonic oxide and sulphur vapor through a red-hot tube. The reactions made to take place are

$$MgO + 2CS^2 + CO^2 = MgS + 3CO + 3S$$
$$3CO + 3S = 3COS.$$

"The product obtained thus contained 58 per cent. MgS and 42 per cent. undecomposed magnesia. In acting on alumina in the same way, the product obtained is a mixture of Al^2S^3 and alumina."

* Jahresb. der Chem., 1867, 155.

Reichel next tried the different methods which have been proposed to reduce these sulphides to metal, and thus records his results (see p. 183) :—

" Petitjean* patented a process in England for reducing Al^2S^3 by hydrogen acting at a high temperature, or by melting it with iron filings. MgS, heated a long time in a current of hydrogen, remained unchanged. I mixed MgS with iron filings, put it in a porcelain crucible, covered with fresh, dry, fine NaCl, and filled the crucible to the rim with carbon. To keep out all oxygen, I put the crucible inside a larger Hessian crucible, filling in between with pulverized charcoal. After heating several hours in a wind furnace, I found a half-sintered mass under the NaCl. This material, on being boiled with water, evolved no trace of hydrogen sulphide but only pure hydrogen. This showed that the iron had taken the sulphur from the MgS. Still, I did not succeed in extracting the free magnesium from it by amalgamation. In the same manner, Al^2S^3 appeared to be decomposed by iron and heat, but it was also impossible in this case to separate the metallic aluminium out of the mass. Copper effects the reduction as well as iron, forming CuS.

" Since magnesium is not sulphurized on ignition in a current of hydrogen sulphide, it appeared probable that MgS might be reduced by ignition in a stream of hydrogen. I first tried a current of illuminating gas, well dried and freed from hydrogen sulphide by a potash tube. In spite of long ignition, the MgS was unaltered. Then I tried hydrogen, but that also was unsuccessful, the MgS would not give up its sulphur to hydrogen at a bright red heat. Since hydrogen alone does not act on MgS, it is hardly

* Dingler, 148, 371.

to be expected that a hydrocarbon can remove any sulphur from it.

"To find how carbonic oxide acted towards MgS, I ignited the latter in a stream of this gas. The magnesium sulphide used contained 56.5 per cent. sulphur and 33.0 per cent. magnesium, or 12.47 per cent. more sulphur than the formula MgS allows. Under these circumstances COS was evolved, recognized by forming barium sulphide and sulphate, when led into baryta water. As soon as these gases ceased coming off, I cooled the tube in a current of carbonic oxide. The material had retained its former color and still readily evolved hydrogen sulphide in moist air or water. But it had lost 12.23 per cent. of its weight. The gas, it appears, had united only with the sulphur in excess of that required to form MgS, and the polysulphide was thus changed to the monosulphide."

Reichel makes the following summary:—

"The above researches show that magnesium and aluminium can unite with sulphur directly at a high temperature. Also, that MgS and Mg^2OS will be formed when magnesia is similarly treated. Alumina is unattacked by sulphur. Alumina and magnesia are changed by ignition in carbon bisulphide to sulphides. When carbon bisulphide and oxide act on magnesia, Mg^2OS remains; alumina is unchanged. Magnesia is changed by ignition in hydrogen sulphide to MgS, but the operation is tedious and imperfect. By melting the oxides with sulphur no sulphides can be obtained; with alumina the contemporaneous action of a reducing agent is necessary, while magnesia melted with carbon and sulphur or heated in hydrogen and sulphur vapor becomes MgS.

"Al^2S^3 possesses a yellow color, is with difficulty fusible, but fuses to a hard crystalline mass. Usually it is obtained as a sintered yellow powder. In damp air or water the following reaction takes place:—

$$Al^2S^3 + 6H^2O = Al^2(OH)^6 + 3H^2S.$$

"It burns in the air to alumina and sulphur dioxide. MgS forms a polysulphide, as we have seen, but Al^2S^3 does not.

"Also, Al^2S^3 and MgS appear to be reduced at a high heat by metals which have a greater affinity for sulphur, yet it remains to be seen whether this property is technically valuable."

Leaving these two experimenters, Fremy and Reichel, we have very few allusions to the subject. Those who have proposed to produce aluminium from Al^2S^3 state merely that they use Fremy's process for preparing the Al^2S^3.

We have found an article[*] in which it is proposed to pass vapor of carbon bisulphide and hydrochloric acid together over ignited alumina, Al^2S^3 being formed as an intermediate product, and Al^2Cl^6 ultimately formed by the action of the acid. The writer states that by passing the first alone over the ignited alumina the gas evolved is mostly COS, though a portion of it is decomposed to sulphur and carbonic oxide. He further states that Al^2S^3 is only slightly acted on by sodium chloride, is unaffected by calcium or magnesium chlorides, slightly acted on by potassium chloride, but readily chloridized by hydrochloric acid.

F. Lauterborn (see p. 206) claims in a patented process that by calcining aluminium fluoride with calcium sulphide

[*] Chem. News, Dec. 19, 1873.

Al^2S^3 results. I cannot find any corroboration of this statement.

Mr. Niewerth's process for reducing aluminium, in which he either uses Al^2S^3 or else makes it as an intermediate product, will be found in full on p. 185, it being too long to repeat here. I cannot find any outside testimony as to the possibility of his schemes.

Reichel has probably proven the possibility of reducing Al^2S^3 by a metal having more affinity for sulphur. From a chemical standpoint its reduction by copper, iron, etc., should be under the proper conditions a very easy operation. These conclusions follow from the relative affinity of sulphur for the metals, which is set forth in the following investigation:—

"A. Orlowsky* has studied the affinity of sulphur for the metals. From his researches it was found that it usually possesses the greatest affinity for the alkaline metals, with which it forms polysulphides. Among the other metals, copper possesses the greatest affinity for sulphur, then follow in order mercury, silver, iron, lead, and after these platinum, chromium, aluminium, and magnesium, whose affinities for sulphur are quite insignificant."

Experiments on Al^2S^3.

Taking the data given in the foregoing papers, I made a series of experiments on first making Al^2S^3 and then on reducing it.

Experiment I.

Took pure, white alumina, made by calcining pure sul-

---

* Jahresb. der Chemie, 1881, p. 24.

ALUMINIUM SULPHIDE. 319

phate of alumina, put it in porcelain boats in a hard glass tube, and passed vapor of carbon bisulphide, CS^2, over it at bright redness for forty-five minutes. The product was cooled out of contact with the air. The result was a grayish-black powder, not sintered together in the least. On analyzing the product by Fremy's method, it showed 12.65 per cent. of Al^2S^3.

Experiment II.

Took equal parts of alumina, sulphur, and charcoal, ground intimately together in a mortar, and served as in Experiment I, prolonging the action of CS^2 to an hour and a half. The product was a grayish-black powder, similar in appearance to the former product. It contained 38.51 per cent. Al^2S^3.

Experiment III.

Repeated Experiment I, but used a porcelain tube, thus allowing a higher heat than the glass tube would stand. The treatment lasted an hour and a half. The product was of similar appearance to the previous ones, and contained 39.54 per cent. Al^2S^3.

Experiment IV.

I placed some ordinary aluminium sulphate, $Al^2(SO^4)^3$.-$18H^2O$, in the porcelain tube, and heated it gradually up to bright redness with the tube open at both ends, calcining it thus for two hours. The result was that the tube was filled with very porous alumina. CS^2 was then passed over it for two hours, the whole being kept at redness. The product was dirty-white, but lemon-yellow in places, and at the yellow parts sintered together. Analyzing an average specimen, it showed 31.16 per cent, Al^2S^3. It is probable that if a yellow piece had been singled out it would have shown much more Al^2S^3 than this average sample.

Experiment V.

I placed some pure alumina in small hollows cut in pieces of charcoal, and placed these in the tube instead of the porcelain boats. The tube was then placed in an assay furnace and heated almost to whiteness for an hour and half, CS^2 being passed through. The product was small, black, fused buttons melted down into the bottoms of the cavities in the charcoal. These lumps were black outside, brittle, compact fracture, and the broken surfaces mottled, dirty-white, and yellow. They had a strong smell of hydrogen sulphide, and when dropped into water this gas was evolved so actively as to make quite a buzz, resembling the action of a piece of zinc dropped into acid. In one or two minutes the button was resolved into a black powder. This product contained 40.43 per cent. Al^2S^3.

Experiment VI.

Repeated Experiment V, but used porcelain boats. The product was still dark, and contained 38.80 per cent. Al^2S^3.

Experiment VII.

Wishing to make a quantity of the substance, I filled the tube with alumina, put it in a hot fire, and passed CS^2 over it three hours. The product was grayish-black, with here and there touches of yellow, with lumps of considerable size sintered together. An average sample of it contained 32.32 per cent. Al^2S^3.

ALUMINIUM SULPHIDE.   321

Tabulating these results we have—

Experiment	I.	II.	III.	IV.	V.	VI.	VII.
Al^2S^3 (p. ct.)	12.65	38.51	39.54	31.16	40.43	39.80	32.32

First I would notice that, as remarked by Fremy, the Al^2S^3 formed incloses the particles of alumina and prevents further action. It seems highly probable that a stirring apparatus to keep the alumina agitated would greatly improve the product. Experiment I gave poor results because the heat was not sufficient; Experiment II was done at a higher heat, with addition of carbon, and Experiment III at a still higher heat, without carbon. It appears from this that the presence of carbon had very little influence on the amount of Al^2S^3 produced. Experiment V, giving the best results, was worked, I think, at a higher heat than any of the others; but Experiment VI was conducted under as nearly as possible the same conditions; however, we may consider the products as being nearly enough alike, the carbon does not appear to have made a marked difference in the product.

To establish such a process on a practical scale, a wrought-iron or fire-clay retort would be necessary, with arrangements to heat it almost to whiteness. Boats of charcoal, holding ample charges of alumina, are made to fit in the retort. Some sort of stirring apparatus to agitate the alumina from time to time should be provided. The CS^2 could be brought in superheated by waste heat from the furnace and passed out into a condenser. Or, to economize still further, the retort might be lengthened, its forepart made a producer of CS^2, by passing sulphur vapor over carbon, and the rear part be filled with the alumina to utilize this CS^2. Many other devices will occur to the practical chemist in running such a process, the above being mere suggestions.

REDUCING THE Al^2S^3.

Experiment VIII.
I took about half a gramme of product of VII, and wrapping it tightly in lead-foil placed it on a cupel and heated in a muffle. Air was kept from the metal by a close-fitting porcelain cover. On removing the lid after a few minutes, there appeared a button of lead with some powder on its surface. I then cupelled the lead at as low a temperature as possible. The metal cupelled away entirely, leaving no aluminium. On repeating with every precaution the result was the same.

Experiment IX.
About one gramme of product VII was wrapped in copper foil, put in a porcelain crucible, and covered with NaCl and a little charcoal. A close cover was put on, the whole placed in the middle of a Hessian crucible, the latter filled up with fine charcoal, and a cover luted on. On heating this an hour at bright redness, hardly whiteness, there resulted a large button of copper. However, its specific gravity was that of pure copper, and a qualitative test showed no trace of aluminium. It occurs to me now that probably the NaCl reacted on the Al^2S^3, forming aluminium chloride and sodium sulphide, preventing the action of the copper.

Experiment X.
Repeated Experiment IX with tinfoil, and heating only twenty minutes. The tin resulting showed some alumin-

ium on a qualitative test, and on analyzing it I found 0.52 per cent. Considering the small amount of sulphide and the rather large amount of tin used, it is probable that nearly all the aluminium present as Al^2S^3 was reduced.

Experiment XI.
Repeated the same, but using powdered antimony to mix with the Al^2S^3. The resulting button was pure antimony with no aluminium in it.

Experiment XII.
Repeated the experiment, employing fine iron filings and using a high heat for one and a half hours. The product was a loose mass in which were small buttons of metal. These buttons were bright, yellower than iron, and contained 9.66 per cent. aluminium.

Lack of time and opportunity prevented my extending these experiments on reduction. I had intended trying copper filings, zinc filings, mercury—excluding air by using a vacuum or an atmosphere of hydrogen—or its reduction by hydrogen gas.

On reviewing the experiments reported above, those with tin and iron succeeded best. Knowing the great affinity of copper for sulphur, I cannot but think that an experiment with very fine copper filings intimately mixed with the Al^2S^3 would give satisfactory results.

In closing I would remark that a process such as suggested on p. 321 could be easily arranged on a large scale, the undecomposed CS^2 being caught and so no more of it used than is necessary to supply sulphur for the Al^2S^3. The product could be mixed with fine metallic filings, put into a crucible, surrounded by charcoal, and the alloy made. The metal changed to sulphide could be recovered by reducing the slags. These processes have been covered by patents, but have never been made successful. It appears that if rightly managed they will give good results and produce aluminium alloys cheaply.

ADDENDA.

Additional Details of Castner's Sodium Process.

"In the ordinary sodium process,* lime is added to the reducing mixture to make the mass refractory, otherwise the alkali would fuse when the charge is highly heated, and separate from the light, infusible carbon. The carbon must be in the proportion to the sodium carbonate as four is to nine, as is found needful in practice, so as to assure each particle of soda in the refractory charge having an excess of carbon directly adjacent or in actual contact. Notwithstanding the well-known fact that sodium is reduced from its oxides at a degree of heat but slightly exceeding the reducing point of zinc oxide, the heat necessary to accomplish reduction by this process and to obtain even one-third of the metal in the charge, closely approaches the melting point of wrought iron.

"In my process, the reducing substance, owing to its composition and gravity, remains below the surface of the molten salt, and is, therefore, in direct contact with the

---

* Journal of the Franklin Institute, Nov. 1886.

fused alkali. The metallic coke of iron and carbon contains about 30 per cent. carbon and 70 per cent. iron, equivalent to the formula FeC^2. I prefer to use caustic soda, on account of its fusibility, and mix with it such quantity of so-called 'carbide' that the carbon contained in the mixture shall not be in excess of the amount theoretically required by the following reaction :—

$$3NaOH + FeC^2 = 3Na + Fe + CO + CO^2 + 3H\ ;$$

or, to every 100 pounds of pure caustic soda, seventy-five pounds of 'carbide,' containing about twenty-two pounds of carbon.

"The necessary cover for the crucible is fixed stationary in each chamber, and from this cover a tube projects into the condenser outside the furnace. The edges of the cover are convex, those of the crucible concave, so that when the crucible is raised into position and held there the tight joint thus made prevents all leaking of gas or vapor. Gas is used as fuel, and the reduction begins towards 1000° C. As the charge is fused, the alkali and reducing material are in direct contact, and this fact, together with the aid rendered the carbon by the fine iron, in withdrawing oxygen from the soda, explains why the reduction is accomplished at a moderate temperature. Furthermore, by reducing from a fused mass, in which the reducing agent remains in suspension, the operation can be carried on in crucibles of large diameter, the reduction taking place at the edges of the mass, where the heat is greatest, the charge flowing thereto from the the centre to take the place of that reduced.

"I am enabled to obtain fully ninety per cent. of the metal in the charge, instead of thirty per cent. as formerly.

The crucibles, after treatment, contain a little carbonate of soda, and all the iron of the 'carbide' still in a fine state of division, together with a small percentage of carbon. These residues are treated with warm water, the solution evaporated to recover the carbonate of soda, while the fine iron is dried, and used over again for 'carbide.'"

New Process for Making Aluminium Chloride.

Mr. Chas. F. Mabery has patented and assigned to Cowles Bros. a new process for making anhydrous aluminium chloride. The patent was granted Oct. 26, 1886. The first claim is for producing it by passing chlorine gas over an alloy of aluminium and some other metal kept in a closed vessel at a temperature sufficient to volatilize the Al^2Cl^6 formed, which is caught in a condenser. The second claim is for passing hydrochloric acid gas through the electric furnace in which alumina is being decomposed by carbon, a condenser being attached as before.

Remarks on the Mitis Castings.

Mr. W. H. Wahl, Secretary of the Franklin Institute, Phila., makes the following remark on this subject :—*

"The simplicity of this process, the certainty with which it can be operated, the uniformity of the product, and its good qualities in respect to strength and ductility, indicate an extended field of usefulness for it. The mitis castings threaten to seriously incommode the manufacturers of malleable castings, for which they not only offer

* Journal of the Institute, Nov. 1886.

a perfect substitute, but one which, in respect to strength and ductility, is distinctly superior, while for many purposes mitis castings can be employed for which malleable castings could not be made. The mitis process has also been applied to the production of steel castings, and with promising results. In one of the methods experimentally tested, the sheet castings were from wrought iron scrap as raw material, with the addition of the proper proportion of cast iron to bring the percentage of carbon to the point required for each special purpose."

PRODUCTION OF ALUMINIUM.

From advance proof-sheets of vol. iii. 'Mineral Resources of the United States,' we learn* that the production of metallic aluminium in the United States increased from 1800 troy ounces in 1884 to 3400 ounces in 1885, valued at $2550. Aluminium bronze, ten per cent., was made to the amount of about 4500 pounds, valued at $1800.

In October, 1886, a Philadelphia instrument maker accepted the offer, from the maker, of a large amount of European aluminium at the price of 50 cents per ounce, the lowest at which aluminium has yet been sold.

* Sci. Am., Nov. 13, 1886.

INDEX.

Academy, Paris, patronage of, towards Deville, 32
Acetic acid, action of, on aluminium, 78
Acid, acetic, action of, on aluminium, 78
   hydrochloric, action of, on aluminium, 75
   muriatic, action of, on aluminium, 75
   nitric, action of, on aluminium, 75
   sulphuric, action of, on aluminium, 74
   tartaric, action of, on aluminium, 78
Acids, organic, action of, on aluminium, 78
Air, action of, on aluminium, 70
Albite, formula of, 43
Alkalies, caustic, action of, on aluminium, 77
Alkaline carbonates, action of, on aluminium, 88
Alloys, aluminium, made by Cowles Bros., 205
   of aluminium, 258–303
     and antimony, 292
     and bismuth, 292
     and boron, 299
     and cadmium, 299
     and carbon, 300
     and copper, 264
     and gallium, 300
     and gold, 298
     and iron, 280
     and lead, 291
     and manganese, 302

Alloys of aluminium and mercury, 261
   and molybdenum, 302
   and nickel, 293
   and nitrogen, 303
   and platinum, 299
   and silicon, 259
   and silver, 295
   and sodium, 303
   and tin, 289
   and titanum, 301
   and tungsten, 302
   and zinc, 287
   with steel, 281
   of brass and aluminium, 276
   of German silver and aluminium, 277
Alumina, composition of, precipitated, 165
   crucibles, use of, for obtaining aluminium, 228
   in purifying aluminium, 243
   extracted from alum stone or shale, 153
   manufacture of, 144–153
   manufactured from cryolite, 146–153
     dry way, 146
     wet way, 152
   precipitation of, by carbonic acid gas, 149
   reduction of, by carbon in presence of copper, 309
   sulphate of, Tilghman's process for decomposing, 144
Aluminate of soda crucibles, use of, in purifying aluminium, 243

INDEX.

Aluminate of soda precipitation by Löwig, 151
   precipitation of, at Salindres, 159, 163
Aluminite, formula of, 44
Aluminium, alloys of, 258–303
   amalgam, properties of, 263
   bronze, solders for, 279
   chemical properties of, 70–89
   Crown Metal Co.'s, 173
      makers of aluminium bronze, 274
   crucibles, use of, for obtaining aluminium, 228
   history of, 25–42
   manufacture at Salindres (Gard), 158–171
   metallurgy of, 90–257
   occurrence in nature, 43–50
   physical properties of, 51–70
   plate as a substitute for tin plate, 247
   protoxide, 26
   reduction of, by other agents than sodium, 180
   silver, 297
   sodium, double chloride, making of, 154–157
   uses of, 243–247
   working in, 235–257
Alum-shale, use of, for making alumina, 153
Alum-stone, use of, for making alumina, 153
Alums, native, impurities in, 45
Alunite, formula of, 44
Amalgam, potassium, used in isolating aluminium, 25
Amalgamation of aluminium, 261, 308
   Wöhler's efforts, 25
American aluminium, 306
   Co., Detroit, 221
   price of, 1883–84, 39
   cryolite, 48
Amfreville-la-mi-Voie, aluminium works at, 29, 33
   process used at, 124
Ammonia, aqua, action of, on aluminium, 78
Analyses of beauxite, 46

Analyses of commercial aluminium, 51, 52
   of Mr. Frishmuth's metal, 307
Analysis of aluminium sulphide, 311
   of Bombay Wootz for aluminium, 283
Animal matters, action of, on aluminium, 87
Animals, aluminium never found in, 44
Annealing of aluminium, 58
Anorthite, formula of, 43
Antimony, alloys of aluminium with, 292
   reduction of aluminium sulphide by, 323
Argentan, a serviceable alloy, 294
Artistic purposes, use of aluminium for, 245

Balance beams, special value of aluminium for, 35
Barattes for precipitating alumina, 163
Barium, alloys with aluminium, 87
   oxide, action on aluminium, 87
Barlow, W. H., on the tensile strength of aluminium, 62
Basset, M. N., reduction of aluminium by zinc, by, 215
Battersea, London, first aluminium works in England, 33
Battery, use of aluminium in the, 75
   to deposit aluminium, 80
Baudrin, P., on a new aluminium alloy, 278
Beating of aluminium, 57
Beaux, analysis of beauxite from, 47
Beauxite, 45–47
   analyses of, 46
   and cryolite, chief source of aluminium, 45
   stimulated production of, 27
   deposits in Ireland, France, etc., 46

INDEX. 331

Beauxite, formula of, 44
  treatment of, at Salindres, 160
  where principally found in France, 160
Bekétoff, on the action of oxide of barium on aluminium, 87
Bell Bros., directions for soldering aluminium, 252
  makers of aluminium at Newcastle-on-Tyne, 33
  stoppage of their aluminium works, 35
Bells, aluminium, 63
Benjamin, Mr., additional details of Castner's sodium process, 141
Benzine, use of, in melting aluminium, 236
Benzon, iron process of, 211
Berlin, aluminium works in, 35
Bertrand, M. A., deposition of aluminium by electricity, 232
Berzelius, investigation on the composition of cryolite, 104
Bessemer converter, reduction of aluminium in the, 207
  reduction of sodium and potassium in, 208
Birmingham, England, Webster's aluminium works at, 36
Bismuth, alloys of aluminium with, 292
Blast furnace, reduction of aluminium in a, 185–204
Books on aluminium, Tissier Bros., 28
  Deville's, 29
  Mierzinski's, 29
Borates, action of, on aluminium, 84
Boron, alloys of aluminium with, 299
Boudaret, M., report on the malleability of aluminium bronze, 267
Bourbouze, M., on an aluminium tin alloy, 290
Brass, aluminium, strength of, 276
  compared with aluminium bronze, 271

Braun, John, deposition of aluminium by electricity, 233
Bremen, aluminium works at, 229
Bromide of aluminium used for making pure aluminium, 243
Bromine, action on aluminium, 88
Bronze, aluminium, 264
  a true alloy, 265
  compared with brass, 271
  compressive strength of, 272
  Cowles Bros., 274
  first exhibited at Paris, 1867, 34
  making of, 267
  malleability of, 267
  manufactured by M. Evrard, 211
  melting point of, 271
  phosphorized, 279
  price in 1878, by the Société Anonyme, 36
  production in the United States in 1885, 327
  specific gravity of, 271
  tenacity of, 266, 267, 270, 276
  silicon-aluminium, 205, 280
  silicon, manufactured by M. Evrard, 211
Brünner, reduction of sodium by, 131
Buchner, G., purification of aluminium from silicon, by, 243
Buff and Wöhler on the solution siliceous aluminium, 260
Bunsen and Deville's electrolytic method of separating aluminium, 41
Bunsen, electrical process of, for depositing aluminium, 222, 225
Burnishing of aluminium, 55

Cadmium, alloys of aluminium with, 299
Caillet, on the amalgamation of aluminium, 261
Calcination furnace for sodium mixture, 133
  Thomson's, for cryolite, 146

332 INDEX.

Calcination retorts for aluminium-sodium chloride at Salindres, 166
Calcutta, aluminium at the exhibition of, in 1883, 246
Calvert and Johnson, making of iron aluminium alloys, 282
— reduction of aluminium by iron, 209
Camden, N. J., aluminium made at, 31
Carbon, action on aluminium, 88
— alloys of aluminium with, 300
— and aluminium, alloy of, 203
— and carbon dioxide, reduction of aluminium by, 187
— as lining for earthen crucibles, 36
— changed to graphite, 193
— dioxide and carbon, reduction of aluminium by, 187
— disulphide, use of, for making aluminium chloride, 317
— for making aluminium sulphide, 310
— reduction of aluminium by, 188–206
Carbonates of alkalies, action on aluminium, 88
Carbonic acid gas, lime-kiln for producing, 150
— used for precipitating aluminium, 149, 163
Carburetted hydrogen, reduction of aluminium by, 182
Casting of aluminium, 237
— its importance, 36
— the largest ever made, 39
Castings, Mitis, alloy used for making, 212
Castner, claims made in his patent, 141
— process for reducing sodium by, 324
— reduction of sodium by, 131
Chalk, object of using, in the reduction of sodium, 133

Chanu, aluminium plant at Rouen, 29
Chapelle, M., on the reduction of aluminium by carbon, 188
Charrière, on soldering aluminium, 248
— use of an aluminium tube in tracheotomy, 87
Chemical classification of aluminium, 88
— properties of aluminium, 70–89
— reactions in the sodium process, 158
Chemically pure aluminium, directions for making, 243
Chlorhydrate of aluminium, 85
Chloride of aluminium, a new process for producing, 326
— Dullo's process for making, 155
— improved method for producing, 157
Chlorides, metallic, action on aluminium, 85
Chlorine, action on aluminium, 88
Christofle, M., castings of aluminium bronze, 265
— gilding of aluminium, 80
— on soldering aluminium, 248
— on the use of aluminium-silver alloy, 296
Classification, chemical, of aluminium, 88
Clay, cryolite, as lining for earthen crucibles, 36
Clays, alumina the base of, 43
Cleaning of tarnished aluminium, 54
Cleveland, manufacture of aluminium in, 41, 190
Coating of metals with aluminium, 231
— iron with aluminium, 247
Coins, use of aluminium for, 247
Color of aluminium, 53
Combinations of aluminium, 43
Combustion of aluminium leaf, 71

INDEX.

Comenge, M., double reaction method of, 184
Commercial aluminium, analyses of, 51
  made chiefly by Deville's process, 41
Compressive strength of aluminium bronze, 272
Condenser for sodium, 131
Conductivity of aluminium for heat, 67
  electric, of aluminium, 66
Converter, the Bessemer, reduction of aluminium in, 207
  reduction of sodium and potassium in, 208
Cooking, aluminium utensils for, 245
  utensils, valuable property of aluminium for, 68
Copper, alloys of aluminium with, 264–280
  deposition of, by aluminium, 81
  freeing of aluminium from, 240
  oxide, action on aluminium, 87
  reduction of aluminium by, 212
  of aluminium-sulphide by, 322
  the quality suitable for aluminium bronze, 267
Coppering of aluminium, 80
Corbelli, of Florence, cyanogen process of, 180
  process of, for depositing aluminium electrolytically, 231
Cornwall, supposed discovery of native aluminium at, 43
Corundum, 49, 50
  discovery in Georgia, 49
  from Georgia, used by Cowles Bros., 205
  price at the mines, 49
  use of, for making aluminium, 37
Cost of aluminium at Salindres in 1872, 172

Cowles's aluminium process, history of, 41
Cowles Bros.' agent in England, 197
  aluminium bronze made by, 274
  aluminous materials used by, 205
  owners of a patent for producing aluminium chloride, 326
  patent claims, 190
  process for the reduction of aluminium by carbon, 189–205
Cross, W., description of American cryolite, 48
Crucible clay, action on aluminium, 84
Crucibles, action of aluminium on siliceous, 259
  alumina, use of, for obtaining aluminium, 228
  in purifying aluminium, 243
  aluminate of soda, use of, in purifying aluminium, 243
  aluminium, use of, for obtaining aluminium, 228
  earthen, action of aluminium on, 35
  iron, used by Rose, 106
  use of, in purifying aluminium, 241
  lime, for melting aluminium, 36
  lining for, 35, 123
  porcelain, use of, for obtaining aluminium, 228
  used in electrolyzing aluminium, 224
Cryolite, Allen Dick's experiments on reduction of, 115
  and Beauxite, chief source of aluminium, 45
  stimulated production of, 27
  Berzelius's investigation of its composition, 104
  clay as lining for earthen crucibles, 36

334 INDEX.

Cryolite, composition of, 119
    decomposition of, 306
        by electricity, 230
    Deville's process for reducing, 119-126
    Dr. Percy's experiments on reducing, 115
    formula of, 44
    general use as a flux, 48
    H. Rose's paper on reduction of, 103-115
    importation of, by the Pennsylvania Salt Co., 48
    imports into the United States, 49
    in the United States, 48
    manufacture of alumina from, 146-153
    occurrence of, 48, 49
    reduction of, 103-129
        at Nanterre, 126
        by ferro-silicum, 207
    Watts's summary of its use, 127
Crystalline form of aluminium, 69
Crystallization of aluminium, 115
Crystallized silicon, 260
Culinary articles, use of aluminium for, 68, 245
Cupellation of aluminium, 71
        from lead, 291
Curaudau, reduction of sodium by, 131
Cyanite, formula of, 44
Cyanogen, reduction of aluminium by, 180

Davy, reduction of sodium by, 131
    unsuccessful efforts to isolate aluminium, 25
Debray, H., aluminium plant at Glacière, 28
Debray, M., statement of, in regard to iron in aluminium, 282
Decomposition furnace for sodium mixture, 134
    of aluminium sulphide by water, 311, 317
Degousse, first successful beater of aluminum leaf, 58

De la Rive, on the action of sulphuric acid on aluminium, 74
Denis, M., of Nancy, remark on the soldering of aluminium, 248
Density of aluminium, 64
Deposition of aluminium by electricity, 255
    by the battery, 80
    electrolytically, 222-234
Deville, aluminium plant at Glacière, 28
    analysis of beauxite by, 47
    book on aluminium, 1859, 29
    charges against Tissier Bros., 28
    conclusion of his book, 30
    description of the reduction of sodium, 132
    experiments on gilding and silvering aluminium, 256
    H. St. Claire, first to isolate pure aluminium, 26
    on soldering of aluminium, 247
    on the aluminium obtained by Wöhler, 95
    on the casting of aluminium, 237
    on the electrolytic reduction of aluminium, 223, 225
    on the melting of aluminium, 235
    on veneering with aluminium, 253
    purification of aluminium, 238
    researches of, at the Normal School, Paris, and at Javel, 28
    on cryolite by, 118
    review of Percy's and Rose's investigations of cryolite, 118
    treatment of silicates and borates with aluminium, 84
Deville's cryolite process, Wöhler's improvement on, 126
    improvements in 1854 for obtaining pure aluminium, 96
    processes, later improvements on, 173

Deville's sodium vapor process, 100
Diaspore, formula of, 44
 where found, 50
Dick, Allan, paper on reducing cryolite, 116
Disthene, reduction of, by an electric current, 229
Donny and Mareska, condenser for making sodium, 131
Double reaction, reduction of aluminium by, 184
Drawing of aluminium into wire, 60
Drechsler, analysis of beauxite by, 47
Dublin, analysis of beauxite from, 47
Ductility of aluminium, 60
Dullo, M., process for making aluminium chloride, 155
 remark on the reduction of aluminium by zinc, 214
Dumas, on gases in aluminium, 52
Duvivier, M., on reduction of aluminium by electricity, 229
Dynamo used in Cowles Bros.' process, 204

Elastic range of aluminium, 62
Elasticity of aluminium, 61
Electric conductivity of aluminium, 66
 furnace, Cowles Bros., 199
 suggested by Miefzinski, 226
 use of, for producing aluminium chloride, 326
Electrical furnace, gases from the, 197
 separation of aluminium, 255
Electricity applied to melting steel, 194
 to the extraction of metals, 194
 reduction of aluminium by, 222
 of sodium by, 131
 unsuccessful efforts of Davy to isolate aluminium by, 25

Electrolytic methods of separating aluminium, 41
Enamelling mixture to protect sodium retorts, 135
England, Cowles Bros.' agent in, 197
 failure of aluminium manufacture in, 35
 first aluminium works in, 33
 Webster perhaps the only maker of aluminium in, 42
Engraving of aluminium, 61
Evrard, M., making of aluminium bronze by, 211

Falk & Co., makers of aluminium leaf, 60
Faraday's experiments on the sonorousness of aluminium, 64
Farmer, Moses G., patented apparatus of, for electrolytically obtaining aluminium, 233
Favre, M., on solution of aluminium in hydrochloric acid, 75
Feisstritz, analysis of beauxite from, 47
Ferro-silicum, reduction of aluminium by, 206
Fixity of aluminium, 66
Fleury, A. L., carburetted hydrogen process of, 182
Fluoride of aluminium, Gerhard's process of reducing, 181
 Lauterborn's process of reducing, 206
 reduction by ferro-silicum, 206
Fluorides as fluxes, 260
 use of, as flux, 102–120
Fluorine, action on aluminium, 88
Fluorspar, action on aluminium, 84
Flux, fluorspar as, 84
 use of fluorides as a, 102–120
Fluxes for aluminium, 259
Formulæ of aluminous minerals, 43
France, deposits of beauxite in, 46

France, production of aluminium in, 1882, 39
   successful manufacture of aluminium in, 35
Fremy, original paper on aluminium sulphide, 310
Frishmuth, aluminium works in Philadelphia, 37
   analyses of the metal of, 306
   improvement of, in making aluminium-sodium double chloride, 154
   plating with aluminium-nickel, 234
   production of aluminium, 1883, 1884, 40
   solders for aluminium by, 250
Frishmuth's first assertions not verified, 42
   patent claims, 178
   owners of, 37
   process mentioned in Watts's Dictionary, 42
   works, annual production of, 39
Fritz, Mr., of Bethlehem, Pa., interest of, in the mitis process, 287
Fusibility of aluminium, 65
Furnace, Thomson's, 146

Gallium, alloys of aluminium with, 300
Garnet, formula of, 44
Gases in aluminium, 52
   from the electrical furnace, 197
Gaudin, electric process of, 230
Gay Lussac, reduction of sodium by, 131
Gerhard & Smith, patent process of, for depositing aluminium, 233
Gerhard, W. F., furnace for reducing aluminium, 127
German silver, aluminium, 277
Germany, aluminium works in, 33
   failure of aluminium manufacture in, 35
   reduction of aluminium in, 228

Gila River, N. M., native sulphate of alumina on the, 305
Gilding of aluminium, 80, 256
Glacière, process for making aluminium, as used at, 98
   purification of aluminium from slag at, 239
   Rousseau Bros., aluminium works at, 28
Glass, action of, on aluminium, 84
Gmelin, an observation on aluminium amalgam, 263
Gold, alloys of aluminium with, 298
Gordon, A., on the tensile strength of aluminium bronze, 266
Gore, deposition of aluminium on copper, 234
Granite, composition of, 43
Graphite, carbon changed to, 193
   cylinders, use of, to protect retorts in sodium reduction, 135
Grätzel, Richard, electrolytic process of, 228
Gravity, specific, of commercial aluminium, 307
Greenland, cryolite in, 48
Grousilliers's improvement of reducing under pressure, 179
Guettier, remarks on aluminium bronze, 273
Guiana, deposits of bauxite in, 46

Hadamar, analysis of bauxite from, 47
Hamburg, aluminium works at, 229
Hardness of aluminium, 61
Harmlessness of aluminium salts to the body, 79
Havrez, P. J., washing apparatus of, 149
Heat, conduction of, by aluminium, 67
   specific, of aluminium, 68
Helmet, an aluminium, 245
Herreshoff, importer of bauxite into the United States, 47

INDEX.

Hesse, analysis of beauxite from, 47
  deposits of beauxite in, 46
Hillebrand, description of American cryolite, 48
Hirzel, on alloys of aluminium and silver, 296
History of aluminium, 25–42
Hodges, F., analysis of beauxite by, 47
Hulot, coppering of aluminium, 80
  method of soldering aluminium, 248
  on the use of aluminium in the battery, 75
Hunt, Dr. T. Sterry, paper on Cowles's process, 194
  second paper on Cowles's process, 196
  views on the aluminium industry, 42
Hydrochloric acid, action of, on aluminium, 75
Hydrogen, action on aluminium, 88
  reduction of aluminium by, 181
  sulphide, action of, on aluminium, 73
  use of, to purify aluminium, 243

Imports of aluminium, 1870 to 1884, 40
Impurities, freeing of aluminium from, 240
Instruments, mathematical, etc., suitability of aluminium bronze for, 268
  optical and portable electric, a suitable alloy for, 290
Iodine, action on aluminium, 88
Ireland, analysis of beauxite from, 47
  deposits of beauxite in, 46
Iron, alloys of aluminium with, 280
  with aluminium, 209
  aluminium alloy, production of, 323

Iron, aluminium alloy, used in the mitis process, 213
  coated with aluminium, 247
  crucibles, use of, in purifying aluminium, 241
  used by Rose, 106
  freeing of aluminium from, 240
  in commercial aluminium, estimation of, 307
  oxide, action on aluminium, 86
  reduction of aluminium by, 206
  sulphide by, 323
Ivigtuk, Greenland, cryolite beds at, 48

Jablochoff, reduction of sodium by, 131
Javel, chemical works at, 28
  process for making aluminium as used at, 98
Jeancon, J. A., patented process for depositing aluminium, 232
Jewelry, aluminium, 247
Johnson, double reaction method of, 184
Jouet, Mr., analysis of beauxite by, 47
Joules, on the amalgamation of aluminium, 262

Kagensbusch, electric process of, 230
  of Leeds, on reduction of aluminium by zinc, 218
  on reduction of aluminium by lead, 222
Kamarsch, on the tensile strength of aluminium, 63
Kerl and Stohman, directions for soldering aluminium by, 250
  historical résumé by, 32
  on the melting of aluminium, 236
Klein-Steinheim, analysis of beauxite from, 47
Knight, remarks on aluminium bronze, 273

INDEX.

Knowles, Sir F. C., cyanogen process of, 180

Lang, I., analysis of bauxite by, 47
Langsdorff, analysis of bauxite from, 47
Lauterborn, iron reduction, process of, 206
Lauterborn's process for decomposing cryolite tested, 306
    remark on, 317
Lavoisier, first to suggest the existence of aluminium, 25
Lazulite, formula of, 44
Lead, action on an aluminium-tin alloy, 196
    alloys of aluminium with, 291
    deposition of, by aluminium, 82
    freeing of aluminium from, 241
    oxide, action on aluminium, 86
    reduction of aluminium by, 221
    sulphide by, 322
Leaf aluminium, 245
    beating of, 57
    combustion of, 71
    decomposition of water by, 73
Lechatelier, M., on the tensile strength of aluminium bronze, 266
Lecoq de Boisbaudran, on aluminium-gallium alloys, 300
"Lessiveur méthodique," 149
Liebig, experiments to reduce aluminium, 93
Lime, action of, on aluminium, 77
    crucibles for melting aluminium, 36
    kiln, use of, to furnish carbonic acid gas, 150
    phosphate of, action on aluminium, 85
Lining for crucibles, 123
Liquation of aluminium, 238–240

Lissajous, M., aluminium tuning fork made by, 63
Litharge, action on aluminium, 86
Lithia mica, formula of, 43
Lockport, N. Y., Cowles Bros., plant at, 193–196
Löwig, experiments of, in precipitating alumina, 151
Lustre of aluminium, 55

Mabery, Prof. Chas. F., official announcement of Cowles Bros.' process, 191–194
    on a new process for producing aluminium chloride, 326
    views on the aluminium industry, 41
Magnesia mica, formula of, 43
Magnesium sulphide, production and reduction of, 312–317
Magnetism of aluminium, 69
Malétra, aluminium plant at Rouen, 29
Malleability of aluminium, 57
    bronze, 267
Mallet, directions for making chemically pure aluminium, 243
    on the resistance of pure aluminium to alkalies, 77
    on the specific heat of pure aluminium, 68
    gravity of aluminium, 65
Manganese, alloys of, with aluminium, 302
    oxide, action on aluminium, 86
    reduction of aluminium by, 222
Manufacture of aluminium bronze, 267
Martin, Wm., aluminium plant at Rouen, 29
Mat, production of, on aluminium, 54
Mayer, L., analysis of bauxite by, 47
Melting of aluminium scraps, 235
    point of aluminium, 65

INDEX. 339

Mercury, action on aluminium, 25
alloys of aluminium with, 261
deposition of, by aluminium, 81
Merle & Co., aluminium works at Salindres, 28, 35
Metallic aluminium not found native, 43
chlorides, action on aluminium, 85
oxides, action on aluminium, 86
Metallurgy of aluminium, 90, 257
general remarks on the, 128
of sodium, 130–143
Metals, coating of, with aluminium, 231
comparative density of, 64
electric conductivity of, 67
thermal conductivity of, 67
plating on, with aluminium, 255
precipitation of, from solution by aluminium, 79
relative affinity of sulphur for, 318
Michel, experiment on aluminium and molybdenum, 302
making of iron-aluminium alloys by, 282
Mierzinski, formulæ of some aluminium minerals as given by, 43
general remarks on alloys of aluminium, 258
on making aluminium bronze, 267
on the manufacture of aluminium-sodium chloride, 154
on the melting of aluminium, 237
on the reduction of aluminium by electricity, 226
remark on the electrolysis of aqueous solutions of aluminium salts, 234

Mierzinski, report on the present state of the alumina industry, 144
Minargent, composition of, and method of making, 278
Mineral soda, 105
Mitis castings, 283
ores suitable for, 205
W. H. Wahl's remarks on, 326
process, iron-aluminium alloy used in the, 212
Molten aluminium, viscidity of, 237
Molybdenum, alloys of, with aluminium, 302
Monnier, Alfred, maker of aluminium at Camden, N. J., 31
Morin and Deville, experiments in gilding and silvering aluminium, 80
Morin, P., aluminium plant at Glacière, 28
experiments on gilding and silvering aluminium, 256
improvements by, at Nanterre, 28
on the action of wine on aluminium, 78
on the specific heat of aluminium, 68
Morris, J., carbon and carbon dioxide, method of, 187
Mourey, method of soldering aluminium, 248
receipt for removing tarnish from aluminium, 54
success in gilding and silvering aluminium, 80, 256
Muriatic acid, action of, on aluminium, 75

Nanterre, aluminium works at, 28, 35
production of aluminium at, 1859, 33
reduction of cryolite at, 126
Napoleon III., liberality of, 28
Native aluminium, 43
sulphate of alumina, 305
Neogen, composition of, 277

Newcastle-on-Tyne, Bell Bros., aluminium works at, 33
New York City, manufacture of sodium in, 139-141
Nickel, alloys of aluminium with, 293
  aluminium plating by Frishmuth, 234
  experiment on aluminium and tungsten, 302
Niewerth, double reaction, method of, 185
  iron reduction, process of, 206
Niewerth's nascent sodium process, 179
  process, remark on, 318
Nitre, action of, on aluminium, 83
  purification of aluminium by, 241
Nitric acid, action of, on aluminium, 75
Nitrogen, action on aluminium, 88
  alloys of, with aluminium, 303
Normal School, Paris, experiments at the, 28

Occurrence of aluminium in nature, 43-50
Odor of aluminium, 56
Oerstedt, first published paper on aluminium, 90
  isolation of aluminium by, 25
Oerstedt's paper reviewed by Wöhler, 91
Ores of aluminium used by Cowles Bros., 205
Organic acids, action of, on aluminium, 78
Orlowsky, A., on the relative affinity of sulphur for the metal, 318
Orthoclase, formula of, 43
Ostberg, Peter, inventor of mitis castings, 283
  remark on the reduction of aluminium by iron, 212
Oxidation of aluminium, 71
Oxide of barium, action on aluminium, 87

Oxide of copper, action on aluminium, 87
  of iron, action on aluminium, 86
  of lead, action on aluminium, 86
  of manganese, action on aluminium, 86
  of zinc, action on aluminium, 86
  sub, of aluminium, 26
Oxides, metallic, action on aluminium, 86

Palais de l'Industrie, Paris, 1855, aluminium bar exhibited at, 34
Paraffin, Wagner's use of, to preserve sodium, 138
Paris Exhibition of 1855, aluminium objects presented at, 28
Passementere, aluminium, 60
Peligot, on cupelling aluminium, 71
Pennsylvania Salt Co., importers of cryolite, 48
Pens, suitability of aluminium bronze for, 272
Percy, Dr., experiments in making aluminium bronze, 264
  reduction of cryolite prior to Rose, 115
Petitjean, carburetted hydrogen, process of, 183
Petitjean's process tested by Reichel, 315
Philadelphia, Col. Wm. Frishmuth's aluminium works in, 37
Phosphate of lime, action on aluminium, 85
Phosphorized aluminium bronze, 279
Phosphorus in aluminium, 126
Photo-salts of aluminium, efforts to produce, 27
Physical properties of aluminium, 51-70
Plants, alumina never found in, 44
Plating, aluminium, 231
  aluminium-nickel, 234
  with aluminium, 246

INDEX. — 341

Platinum, alloys of aluminium with, 299
Poggendorff and Reiss on the magnetism of aluminium, 69
Polish of aluminium, 55
Porcelain crucibles, use of, for obtaining aluminium, 228
imitation of, made with cryolite, 48
Potash, action of, on aluminium, 77
mica, formula of, 43
Potassium, aluminium first isolated by the use of, 26
amalgam, experiment with, by Gmelin, 263
used in isolating aluminium, 25
and sodium, reduction together of, 138
carbonates, action on aluminium, 88
chloride, action on aluminium, 85
decomposition of, by electricity, 142
cyanide as a reducing agent, 180
reduction in the Bessemer converter, 208
in the electric furnace, 193
replaced by sodium by Deville, 27
sulphate, action on aluminium, 88
vapor of, Davy's efforts to isolate aluminium with, 25
Precious stones, formulæ of, 44
Precipitation of alumina at Salindres, 163
of metals from solution by aluminium, 79
Price of aluminium in 1857, 29
in 1878, by the Socié.é Anonyme, 36
in 1883–84, 39
in October, 1886, 327
Proctor, Bernard S., comparison of brass with aluminium bronze by, 271

Production of aluminium bronze in the United States in 1885, 327
by Col. Frishmuth, 1883–1884, 40
in France, 1882, 39
in the United States in 1885, 327
Properties of aluminium sulphide, 311, 316
Pure aluminium, chemically, directions for making, 243
only made by using sodium, 42
requirement for making, 205
Purification of aluminium, 74, 238
by nitre, 83

Rammelsberg, on silicon in commericial aluminium, 52
Rammelsberg, Prof., experiments in reducing cryolite, 103
Rattle, baby, first article made of aluminium, 244
Reduction furnace, Deville's, for sodium, 134
for reducing aluminium by sodium, 169
Gerhard's, 127
of alumina by carbon in presence of copper, 309
of aluminium at Salindres, 168
by carbon, 188
and carbon dioxide, 187
by carburetted hydrogen, 182
by copper, 212
by cyanogen, 180
by double reaction, 184
by electricity, 222
by ferro-silicum, 206
by hydrogen, 181
by iron, 206
by lead, 221
by manganese, 222
by other agents than sodium, 180

29*

Reduction of aluminium by silicon, 207
  by zinc, 214
  sulphide, 315, 322
  of sodium by Castner, 324
  under pressure, 179
Reflectors, advantages of aluminium for, 245
Regnault, M., on the specific heat of aluminium, 68
Reichel, paper on sulphides of aluminium and magnesium, 312–317
Reinar, G. W., on the reduction of aluminium by carbon, 189
Reiss and Poggendorff, on the magnetism of aluminium, 69
Retorts for reducing sodium, 134
Retzlaff, analysis of beauxite by, 47
Ricarde-Seaver. Major, views on aluminium, 37
Rolling of aluminium, 57
Rose, H., paper on reduction of cryolite, 103–115
Rouen, aluminium works near, 29, 33
  process used at, 124
Rousseau Bros., aluminium works at Glacière, 28
Ruby, formula of, 44

St. Austel, supposed discovery of native aluminium at, 43
Salindres, aluminium works at, 28, 35
  cost of aluminium at, 172
  manufacture of aluminium at, 158–174
  of aluminium-sodium chloride at, 154–166
Salt, common, action on aluminium, 85
Salts, metallic, action of solutions of, on aluminium, 79
Sapphire, formula of, 44
Sartorius of Göttingen, first maker of aluminium balance beams, 35
Sauvage, F. H., inventor of neogen, 277
Schank, washing apparatus of, 149
Schnitzer, analysis of beauxite by, 47
Schwarz, improvement on Mourey's solders by, 249
Scraps, aluminium, melting of, 235
  melting of, by Col. Frishmuth, 39
Sellers, Mr., of Philadelphia, on the use of aluminium in casting iron, 284
Senet, M. L., on depositing aluminium by electricity, 233
Sevrard, M., success of, in veneering aluminium, 253
Seymour, Fred. J., patent for the reduction of aluminium by zinc, 218
  second patent of, 220
Shaw, T., patented phosphor aluminium bronze, 279
Siemens's furnace, used in reducing sodium, 135
Siemens, Sir Wm., melting of steel with electricity, 194
Silicates, action of, on aluminium, 84
  of aluminium, formulæ of, 43
Siliceous aluminium, 238
  on the gases evolved during solution of, 260
Silicon, alloys of aluminium with, 259
  aluminium bronze, 205
  extraordinary strength of, 280
  bronze manufactured by M. Evrard, 211
  crystallized, 260
  disengagement of, as silicuretted hydrogen in dissolving aluminium, 76
  facilitates the oxidation of aluminium, 71
  freeing of aluminium from, 243
  its state of combination in aluminium, 52
  use of, to reduce aluminium, 207

INDEX. 343

Silicuretted hydrogen, formation of, on dissolving aluminium, 260
Silver, alloys of aluminium with, 295
 aluminium, 297
 comparative value with aluminium, 65
 deposition of, by aluminium, 81
 from clay, exhibited at Paris, 1855, 34
 sulphide, decomposition of, by aluminium, 74
Silvering of aluminium, 256
 difficulty in, 80
Smith, Dr., patents on reduction of aluminium, 221
Société Anonyme de l'aluminium, 35
 prices of aluminium and aluminium bronze, 36
Soda, action of, on aluminium, 77
 mica, formula of, 43
 mineral, 105
Sodium, alloys of, with aluminium, 303
 aluminate, precipitation of, by Löwig, 151
 amount necessary to reduce aluminium from cryolite, 122
 and potassium, reduction together of, 138
 calcination furnace, 133
 carbonates, action on aluminium, 88
 chloride, action on aluminium, 85
 decomposition of, by electricity, 142
 cyanide as a reducing agent, 180
 Gerhard's furnace to prevent loss of, 126
 great reduction of its price in 1859, 27
 its manufacture, 130–143
 manufacture in New York City, 1886, 139–141
 mixture for reduction, 132

Sodium, nascent, as a reducing agent, 179
 preservation of, by Wagner's method, 138
 reaction for the reduction of, 137
 reduction by Brünner, 131
  by Castner, 139
   additional details of, 324
  by Curaudau, 131
  by Davy, 131
  by electricity, 142
  by Gay Lussac, 131
  by Jablochoff, 142
  by Thenard, 131
  furnace for, 134
  in the Bessemer converter, 208
  in the electric furnace, 193
  of aluminium by other agents than, 180
  of the double chloride by, 168
  process, the perfection of the, 171
  substituted for potassium by Deville, 27
 sulphate, action on aluminium, 88
 temperature of the reduction of, 137
 use of chalk in the reduction of, 133
 vapor process as used by Deville, 100
  of Frishmuth, 178
 Weldon's calculation of the cost of, 139
Soils, alumina the base of, 43
Soldering liquor for aluminium, 250
 of aluminium, 247–253
Solders for aluminium, 248–251
 bronze, 279
Sonorousness of aluminium, 63
Spear, Mr. W. B., in connection with native sulphate of alumina, 305
Specht, on the reduction of aluminium by zinc, 218

INDEX.

Specific heat of aluminium, 68
  gravity of aluminium, 64
    bronze, 271
    of commercial aluminium, 307
Sprague, remarks on electrolysis of aluminium salts, 232
Spruce, Mr., analysis of beauxite by, 47
Stamping of aluminium, 58
Stearic acid used for burnishing aluminium, 55
Steel, alloys with aluminium, 281
Stocker, on native aluminium, 43
Stoddart and Faraday, analysis of Bombay wootz, 283
Stones, precious, formulæ of, 44
Strange, Mr., experiments with aluminium bronze, 273
Strength, compressive, of aluminium, 62
  aluminium bronze, 272
  tensile, of aluminium, 62, 63
    of aluminium bronze, 266, 267, 270, 276
    of aluminium-silicon bronze, 205
  transverse, of aluminium, 62
Sulphate of alumina, decomposition of, by electricity, 231, 233
  native, 305
  Tilghman's process for decomposing, 144
  of potash, action on aluminium, 88
  of soda, action on aluminium, 88
Sulphide appearance, properties and analysis of, 311
  a practical process for producing, 321
  experiments on producing, 318
  on reducing, 322
  production and reduction of, 309-324
  reduction of, by Comeuge's method, 184
  by manganese, 222

Sulphide appearance, reduction of, by Petitjean's method, 183
  remark by Than on, 314
  researches of Fremy on, 310
  of Reichel, 312
  carbon di-, use of, for making aluminium chloride, 317
  use of, for making aluminium sulphide, 310
Sulphides of aluminium and magnesium, Reichel's paper on, 312-317
Sulphur, action on aluminium, 73, 88
  its relative affinity for the metals, 318
Sulphuretted hydrogen, action of, on aluminium, 73
Sulphuric acid, action of, on aluminium, 74
Surgery, use of aluminium in, 87
Sweat, action on aluminium, 87
Syndicate, an English, to control patents on aluminium, 39

Tanks, precipitating, 152
Tarnish, Mourey's receipt for removing, from aluminium, 54
Tarnishing of aluminium, cause of, 240
Tartaric acid, action of, on aluminium, 78
Taste of aluminium, 57
Taylor, W. J., calculated cost of aluminium, 32
Telegraph wire, aluminium, 243
Temperature necessary to reduce sodium, 137
Tempering of aluminium, 58
Tenacity of aluminium, 61
Tensile strength of aluminium, 62, 63
  of aluminium bronze, 266, 267, 270, 276
  of aluminium-silicon bronze, 205
Than, remark by, on the formation of aluminium sulphide, 314

INDEX. 345

Thenard, reduction of sodium by, 131
Thermal, conductivity of aluminium, 67
Thomas and Tilly, process for aluminium plating, 231
Thompson, J. B., deposition of aluminium from solution, 231
Thompson, W. P., paper on the Cowles's process, 197–205
Thompson, W. P., reduction of aluminium by, 207
Thomson's calcination furnace for cryolite, 146
"Tiers Argent," an alloy of aluminium and silver, 297
Tilghman's process for decomposing sulphate of alumina, 144
Tin, alloy of aluminium and, 196
    alloys of aluminium with, 289
    aluminium alloy, action of lead on, 196
    production of, 322
    its injurious effects in food, 79
    plate, aluminium plate a substitute for, 247
    reduction of aluminium sulphide by, 322
Titanum, alloys of aluminium with, 301
Tissier Bros., Deville's charges against, 28
    experiments on soldering aluminium, 248
    history of their works, 29
    process, 124
    "Recherches sur l'Aluminium," 1858, 28
Tissier, on the amalgamation of aluminium, 262
Topaz, formula of, 44
Tracheotomy, aluminium tube used in, 87
Tungsten, alloys of aluminium with, 302
Tuning forks, aluminium, 63
Turquois, formula of, 44

Uses of aluminium, 243–247

Usiglio, manager of the works at Salindres, 33

Vanadium, occurrence in bauxite, 46
Vangeois, first maker of aluminium wire, 60
Veneering with aluminium, 253
Vielle Montagne Zinc Works, on the use of retorts from, 288
Viscidity of molten aluminium, 237
Volatilization of aluminium, 66

Wagner, O., analysis of bauxite by, 47
Wahl, W. H., remarks on mitis castings, 326
Wasellite, formula of, 44
Washing apparatus, 149
    used at Salindres, 161
    methodical, 149
Washington monument, cast-aluminium tip of, 39
    composition of the aluminium in the tip of, 53
    description of the apex of, 40
Water, action of, on aluminium, 72
    sulphide, 311, 317
    decomposition of, by aluminium leaf, 73
Watts's Dictionary, Frishmuth's process mentioned in, 42
Watts, on the use of cryolite for producing aluminium, 127
Webster, aluminium works of, at Birmingham, 36
    improvement of, in making aluminium-sodium double chloride, 154
    James, patented alloy of, 278
    process of, only one used in England, 42
Webster's patent, 175
    process, 173–177
Wedding, M., remarks on Basset's zinc process, 217
Weldon, W., of Burstow, England, manganese process of, 222

Wertheim, on the elasticity of aluminium, 61
Wilde, A. E., on reducing aluminium by lead, 221
Winckler, Dr. Clemens, historical retrospect to 1879, 33
— on coating metals with aluminium, 254
— remarks on electrolysis of aluminium salts, 232
Wine, acid action of, on aluminium, 78
Wire, aluminium, drawing of, 60
— strength of, 63
Wirz & Co., Berlin, stoppage of their aluminium works, 35
Wochein, analysis of beauxite from, 47
Wocheinite, local name for beauxite, 46
Wöhler and Buff, on the solution of siliceous aluminium, 260
— and Deville, on the extraction of Boron, 300
— Deville's opinion of, 31
— discovery of aluminium, 1827, 26, 91
— experiments to obtain aluminium amalgam, 25
— first paper of, 91
— improvement on Deville's cryolite process, 126
— method used in 1845, 26
— observation on melting aluminium with a blowpipe, 71

Wöhler and Buff on the resistance of aluminium to aqua ammonia, 78
— review of Oerstedt's paper, 91
— second paper of, 93
Working of aluminium, 235-257
Works, aluminium, at Amfreville, near Rouen, 29, 124, 128
— at Battersea, London, 33
— at Berlin, 35
— at Birmingham, England, 36, 173
— at Camden, N. J., 31
— at Cleveland, Ohio, 41, 189
— at Glacière, 28, 98
— at Javel, 28, 98
— at Nanterre, 28, 35, 126, 128
— at Newcastle-on-Tyne, 33
— at Philadelphia, 37, 178
— at Salindres, 28, 33, 128, 168
— in England, 33, 35, 42
— in France, 35, 128
— in Germany, 33, 35

Zinc, alloys of aluminium with, 287
— deposition of, by aluminium, 83
— expelling from aluminium by heat, 218
— freeing of aluminium from, 242
— oxide, action on aluminium, 86
— reduction of aluminium by, 214

www.ingramcontent.com/pod-product-compliance
Lightning Source LLC
Chambersburg PA
CBHW030325240426
43673CB00040B/1276